Fall Color and
Woodland Harvests

C. Ritchie Bell
Anne H. Lindsey

Fall Color and Woodland Harvests

A Guide to the More Colorful Fall Leaves and Fruits of the Eastern Forests

Laurel Hill Press
Chapel Hill, North Carolina

Library of Congress Catalog Card Number: 90-91636

Frontispiece: Dogwood, Chapel Hill, NC

Design by Kachergis Book Design, Pittsboro, NC

Printed and bound in Singapore by Tien Wah Press

Contents

Acknowledgments

This book is an outgrowth of the authors' development of a public service program of the North Carolina Botanical Garden, first offered at High Hampton Inn in Cashiers, NC, in October 1982. Our acknowledgments for much direct and indirect help with the course, and with ideas and materials for this book, go back to that date.

We acknowledge with many thanks the initial and continuing role in the course (and the influence on this book) of James Ward, Dorothy Wilbur, Kenneth Moore, and Charlotte Jones-Roe of the North Carolina Botanical Garden staff, and the excellent "on site" help of Inas Crisp of the High Hampton Inn staff with all aspects of the program. Special appreciation must be expressed for the personal hospitality of William McKee, his son Will, and all of his staff at High Hampton. Their many kindnesses have contributed greatly not only to the success of the Garden's Fall Color and Woodland Harvest program, but to the authors' enjoyment of their seasonal work on this book while in the mountains of western North Carolina.

A number of friends and colleagues have provided specific bits of information for, and helpful comments on, the manuscript. Dr. Albert E. Radford kindly provided identification, or confirmation of identifications, on several of the photographs included.

No photographer can hope to cover every area of spectacular fall color in a 200-mile-wide swath from northern Georgia to the St. Lawrence River, even over a period of years. Thus the scope of this book has been made possible only by drawing on the work of other fall color photographers who have generously made their work available for inclusion. To these people, whose names, locations, and identifying initials are given on page 16, we are grateful beyond words. Their initials in the text identify the beautiful photographs they supplied. Because we sometimes dipped rather heavily into the photographs of a few people, an extra word of appreciation should go to Barbara Hallowell, Hugh Morton, David Ramsey, Fred Swope, John Weaver, and Peter White.

Thanks are due, and gladly tendered, to Dwight Kline for his friendly assistance (often provided on short notice!) with the several hundred color prints needed by the authors in the initial organization and layout of this book; and to Hiden Ramsey for making available the meteorological data used in the daylight and temperature charts. We thank Jess Bell for his superb job of editing the manuscript.

Of course, half the fun of doing a book is to work with Joyce Kachergis, her daughter Anne Theilgard, and their talented group of design and production people. This is our third —we may try for four!

Fall Color and Woodland Harvests

A [RS]

Why We Have Fall Color

As the days grow shorter toward the end of summer, the stage is set for the biological equivalent of the "Greatest Show on Earth," the annual change in color of the eastern deciduous forests from summer green to the brilliant hues of fall. Actually, the show really starts, as it has for thousands of years, on the longest day of the year and the time of the summer solstice, the first day of summer, June 21. That is the day when the tilt of the earth begins to change and the sun "starts back south," causing the annual changes in light and temperature (see following tables) that give us the four seasons. With the shorter days and less direct sunlight the weather becomes cooler, and this change in temperature increases the tempo of internal changes in the plants that will soon produce the brilliant leaf colors in the trees, shrubs, and vines of our deciduous forests.

Although the chemical changes that turn leaves from pale yellow-green in spring to dark green in summer and then to yellow or red in the fall (Figures B, C, D) involve complex biochemical reactions, these changes can be summarized rather briefly. The green leaves of summer owe their color to the green pigment chlorophyll. However, the leaves also contain large amounts of the yellow pigments carotin and xanthophyll. Because the darker green of the more abundant chlorophyll masks the lighter color of the yellow, in the summer the leaves appear green. With shorter days and cooler weather signaling the approaching winter, the plants stop producing chlorophyll (which they have been doing all summer). Then, as the chlorophyll still in the leaves breaks down into its colorless components, the more stable yellow pigments remain and the leaves "turn yellow."

In many other trees, such as Sassafras, Blackgum, and some of the maples, sugars and other compounds in the leaf are converted into other compounds, called anthocyanins, that are red. As the chlorophyll in these leaves breaks down they turn orange, bright red, or maroon depending on the mixture of red and yellow pigments. In the oaks and a few other trees which have leaves with a high tannin content, the leaves may just turn brown as the chlorophyll breaks down. Of course, the red and yellow pigments also soon break down and all leaves ultimately turn brown, after which they rot and their minerals return to the soil.

It is the combination of various amounts of red, yellow, and brown in the leaves of the many different species of trees that gives our forests their great array of spectacular fall color.

Although individual trees, individual leaves of a single tree, or even parts of a particular leaf of some trees such as Red Maple, Sugar Maple, and Sassafras may be green, yellow, red, or brown, or any combination of these colors at any one time, many species of woody plants have a generally characteristic fall leaf color. The birches, aspens, and Tulip Poplar are typically a clear yellow, Beech a bronze-yellow, and the hickories a golden yellow. Tulip Poplar, birches, Beech, and the hickories never have red leaves. On the other hand, the autumn leaves of Dogwood, Sourwood, Sumac, and Blackgum are usually various shades of red. In these species with predominantly red leaves yellow leaves may occur under certain conditions, but many yellow-leaved species do not seem to have the chemistry to produce red leaves.

Throughout the biological world, color and color pattern are critical to the survival of both individual organisms and the species as a whole. In animals color and pattern often function as camouflage for both predator and prey, or as warning color, or as mimicry, or as part of the mating ritual. In plants flower color is important in attracting the pollinators necessary for seed production. Fruit color, and sometimes seed color, play a primary role in the distribution of the seeds of many plants by birds. However, the brilliant array of fall colors in our deciduous woodlands occurs in only a few other places on earth and does not seem to have any direct biological value or function. Rather it appears to be only a most beautiful and serendipitous byproduct of the autumn metabolism of the forest plants.

B
C

E [BGH] F

Why and Where Trees Go Dormant

In mild, uniform tropical and subtropical climates with adequate rainfall, there are green leaves on the trees all the time and individual leaves die and are shed throughout the year. The only tropical trees that periodically shed all their leaves and go dormant are those in areas with distinct seasons based on rainfall. As the dry season begins each year, the leaves, which lose a lot of water through evaporation, begin to dry out, die, and drop off. The trees live through the remainder of the dry season in a dormant state. When the wet season starts, the trees produce a new crop of leaves, bloom, and set seeds before the next dry season dormant period.

In colder climates very low winter temperatures dry the air and turn all surface water needed by plants to ice; in effect, the plants are in a dry season that puts the same water stress on them as would a desert. Also, at temperatures near or below freezing, photosynthesis, the food-manufacturing process of green plants on which all life on our planet depends for food, stops or becomes very inefficient. Most of the trees in our north temperate forests have adapted to low seasonal temperatures, dryness, and reduced light by dormancy; but both the evergreen conifers and the broad-leaved evergreens have adapted over the centuries to the cold and dryness of winter by biochemical and structural changes (such as thicker leaves and needles) that reduce water loss.

The Evergreen Background

Because evergreens remain green throughout the year, the bright yellows and reds of the autumn leaves of deciduous trees often appear even more brilliant when they are seen against the dark background of their evergreen neighbors. Also an occasional evergreen or patch of evergreens in an otherwise bright red and yellow landscape offers an interesting contrast.

Even evergreens shed some of their leaves on a seasonal schedule as their older leaves become shaded by newer leaves and less efficient in food production. For a plant to be truly evergreen most of its leaves must last at least one full year (from spring to the following spring) or a year and a half (from spring to the fall of the following year). Sometimes, as with Rhododendron, the individual leaves may stay on the plant for five years or more and only drop when they become injured or nonproductive. Before the leaves of an evergreen plant fall, they typically turn yellow or brown (Figures E, F) like those of deciduous trees. However, since the colored leaves or needles are behind the green, new-growth leaves or needles (or if only single leaves are involved), the color is not noticed.

In Canada, in the New England states, and along the crests of the Appalachian Mountains as far south as North Carolina, the evergreen background is often composed of spruce, fir, or hemlock. At lower elevations, and in the southern part of the range, the evergreens are more likely to be various species of pine or cedar or the broad-leaved evergreen American Holly and, along the southeastern coastal plain, Southern Magnolia. Also two very attractive evergreen shrubs, Rhododendron and Mountain Laurel, often form dense stands of dark green foliage in much of the southern Appalachian area. Those who explore the fall woodlands more closely will also find a few evergreen vines, ferns, and low herbaceous plants.

G [BGH]

Where to Find Fall Color

Although the crisp reds and yellows of fall color that cover thousands of square miles in eastern North America involve literally millions of individual trees, only a few species of trees provide the dominant color in any particular area. Thus, even though the forest colors may not change noticeably from one scenic area to the next, different tree species may be involved. For example, in the White, Green, and Adirondack Mountains, most of the autumn yellow will be from the birches, aspens, and Sugar Maple (Figure S); further south, in the Cumberlands and Monongahelas, the yellow is primarily Sugar Maple with some hickory and Tulip Poplar; and still further south, as one approaches the end of the Blue Ridge Parkway, the yellow is primarily hickory and Tulip Poplar (#43). Even so, most of the fall color is provided by fewer than two dozen common and widespread species: three or four species each of maple, birch, oak, and hickory, two of ash, plus Tulip Poplar, Blackgum, and Dogwood. Their color, of course, is supplemented by the colors of various local species that occur in smaller numbers.

Because these colorful native trees (and the trees of their evergreen background) are so widespread and generally attractive, they are also commonly used as landscape plants in yards, parks, and gardens and along the streets of our towns and cities. As a result, everyone in the prime fall color belt is most likely within a few blocks of a number of maples, birches, Tulip Poplars, Blackgums, Beeches, or Dogwoods. Fall color, which can be just as intense on an individual tree in the city as in the forest, is thus only a short walk away for a large part of our population.

To see larger expanses of woodland color may require a short hike or drive out into the countryside. Roads are a boon to fall color because they provide miles of forest edges; the greatest variety of trees (and the best growth and color!) is found at these forest edges, where the trees are less crowded and get more light. In mature forests the edges or "ecotones" are along large streams and rivers, or surround clearings caused by fires, floods, or other catastrophes. Today new edges are created every time a new driveway, road, or highway cuts through a stand of trees; as the trees along the new ecotone grow and branch in response to the additional light, they become more colorful in the fall. In addition new cuts and fills associated with road construction provide new, open habitats for colonization; some of the most intense fall color is produced by young saplings of maple, cherry, Sourwood, and Dogwood growing in full sun on road banks. Indeed, whereas mature trees of Box Elder, for example, have little if any fall color, Box Elder

H

I [PSW]

J

K

L [CAK]©

seedlings and stump sprouts along highways and power line clearings are a beautiful clear yellow (see #62).

At the lower elevations of the coastal plain — especially as one goes further south, where seasonal changes in day length and temperature are less pronounced — fall colors tend to be more muted but can still be quite spectacular (Figure L). And, of course, the more uniform topography of the coastal plain often limits one's view of fall color to a relatively few trees in the immediate foreground. It is inland, in the rolling foothills and the high mountains of the Appalachians and the Alleghenies that form the spine of eastern North America from northern Georgia to southern Quebec, that one finds the vast vistas of great ridges and valleys filled with color.

Most people in the general fall color zone are only a few hours' drive from a peak, ridge, notch, gap, or overlook that affords both a spectacular view of fall color in general and an opportunity to see at close range (and perhaps even identify) some of the trees and shrubs involved in the show.

M [HM]

N [BGH]

Appalachian Area
with major mountain ranges

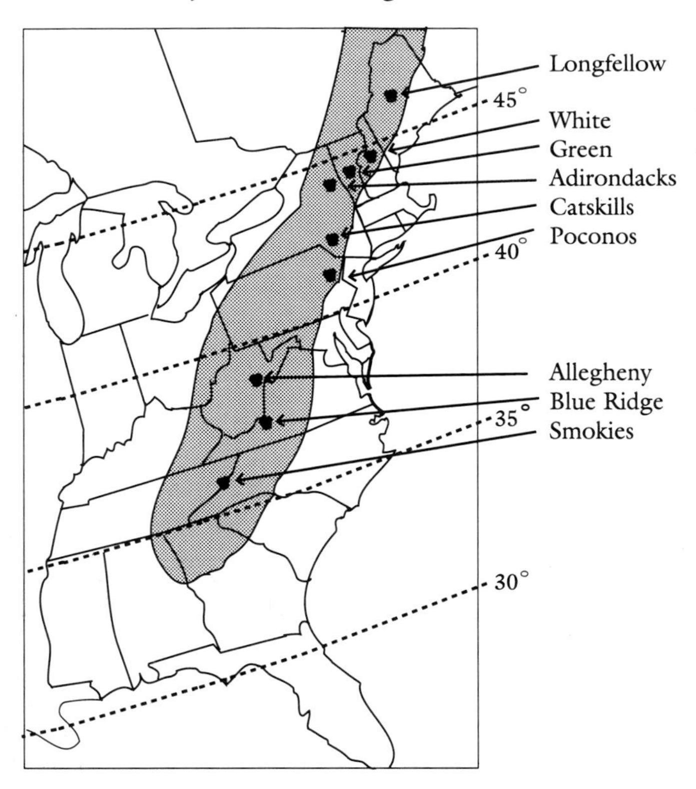

Mountain range	Approx °N latitude	Hours of sunlight		Ave. temp, (°F)	
		21 Jun	21 Dec	21 Jun	21 Dec
Longfellow	46°	16:21	10:47	62	46
White	44°				
Green	44°				
Adirondacks	44°				
Catskills	42°				
Poconos	41°	15:00	11:11	72	68
Allegheny	39°				
Blue Ridge	37°				
Smokies	36°	14:00	12:11	76	73

O [HM]

Fall color starts in the Longfellow Mountains of Maine, the White Mountains of New Hampshire, and the Green Mountains of Vermont about the middle of September and reaches its peak near the end of September or the first week of October, depending on seasonal variations in temperature and local moisture. The Adirondacks, at the same approximate latitude of 44°N, are on essentially the same schedule. About three weeks later, after passing through the Catskills, the Poconos, the Cumberlands, the Monongahelas, and the Blue Ridge, peak color reaches the slopes of the Smokies, at 35°N, some 620 map miles south of the White Mountains but nearly twice that far by road. At the end of October the southern piedmont area gets the last splash of fall color, and by early November the show is over and the eastern forests are dormant. In early spring the longer days and warmer temperatures will cause the leaf buds to swell, and the cycle will start again.

An enjoyable fall foliage trip can be anything from a lunch break visit to see the trees in a nearby park to a month-long adventure from New England to the Smokies, moving south with the color at about 60 miles (by road) each day. Some general sources of helpful information regarding fall color in specific areas are given in Appendix A to help you plan your trip!

Descriptive Format

This book has been written and designed to be as interesting and informative as possible to the many people who enjoy fall color, but with the realization that different people may use it in different ways. If the details do not interest you, just enjoy the pictures! On the other hand, if you want a name or a bit more information on the range, habitat, or uses of a particular plant, this can be found in the descriptive material.

References to text figures are by letter, e.g. "(Figure A)." References to figures associated with specific entries are given by entry number, e.g. "(#43)," or "(#43a)" if there is more than one photograph per entry.

There are two parts: leaves, entries #1–#100, and fruits, entries #101–#147. Each of the 100 entries for leaves has the following information:

1. Entry number and common name
2. Scientific name (genus and species)
3. Comment on leaf length to give scale to the photograph
4. Comment on related species where appropriate
5. General habitat information
6. Notation if plant is not native
7. General distribution (also see map for quick reference)
8. Some plant uses; historical notes where appropriate
9. Photographer's initials (if other than author's)
10. Five-part leaf character code
11. Common and scientific plant family name
12. Reference scale, 2 cm
13. Scaled leaf shape silhouette; multiply by factor "x" to get full size.

Because of space limitations, only items 1, 2, 5, 7, and 9 are supplied for the fruit entries. However, when these fruits are from plants also in the Leaves section, cross-references are given where appropriate.

Comments on the format:

#1. *Common name.* A particular plant species can have only one valid scientific name, but it may have any number of common names. Only the most widespread common name is used here with the entry number; a second common name may be given in the body of the text where appropriate.

#3. *Leaf length.* Since some older, or more general, plant treatments use inches and feet, approximate equivalents in inches and feet are given along with the metric measurements for leaf length, tree height, and land mass elevation.

P
[HM]

#7. *Distribution.* Individual plants, small colonies, or large pure stands of a given species of tree, shrub, or vine usually occur at scattered and often widely spaced localities within the general range of the species. The occurrence of a particular plant in a particular area depends on interactions between the plant and specific local environmental factors such as soil type, moisture, slope, altitude, and other plant species growing in the area. Few plants other than some of our aggressive weeds (such as poison ivy) are really found "everywhere." So, to find a given plant, find the right habitat and keep looking!

#9. *Photograph credits.* We are grateful to the friendly and helpful individuals (and their organizations) noted below. Their initials identify the beautiful photographs they supplied.

RKG	Rob K. Gardner, *North Carolina Botanical Garden, Chapel Hill, NC*
WHG	William H. Gensel, *Chapel Hill, NC*
BGH	Barbara G. Hallowell, *Hendersonville, NC*
WSJ	William S. Justice, *via slide collection donated to the North Carolina Botanical Garden, Chapel Hill, NC*
CAK	Cynthia and Amor Klotzbach, *Cherry Hill, NJ*
HM	Hugh Morton, *Grandfather Mountain, Linville, NC*
LO	Lowell Orbison, *Asheville, NC*
JDP	J. Dan Pittillo, *Western Carolina University, Cullowhee, NC*
GCP	George C. Pyne, *Durham, NC*
RVP	Rick Vande Poll, *Antioch (N.E.), Keene, NH*
WTRP	Wintergreen Team Russell Photography, *Wintergreen, VA*
DHR	David H. Ramsey, *Charlotte, NC*
GSR	Gerald S. Ratliff, *West Virginia Department of Natural Resources, Charleston, WV*
SR	Sam Ristich, *Cumberland, ME*
FCS	Fred C. Swope, *Virginia Military Institute, Lexington, VA*
SJS	Stephen J. Shaluta, Jr., *West Virginia Department of Commerce, Charleston, WV*
RS	Ron Snow, *West Virginia Department of Commerce, Charleston, WV*
JCW	John C. Weaver, *Rancho Palos Verdes, CA, via the Vermont Travel Division, Montpelier, VT*
PSW	Peter S. White, *North Carolina Botanical Garden, Chapel Hill, NC*

Q [GSR]

R [RKG]

#10. *Leaf character summary code.* Plants are usually classified on the basis of a series of more or less technical flower and fruit characteristics. However, many of our tree species can be identified by a set of relatively nontechnical leaf, bark, and fruit characters; indeed, quite a number can be correctly identified by leaf characters alone. The brief five-part leaf character code is a general summary of five important plant and leaf characters needed, in addition to leaf size, for accurate leaf identification of the 100 woody species that account for most fall color.

The letters of the code come from, and represent, specific characters noted in the small illustrated glossary inside the back cover of the book. Thus the code T-A/SCS for Quaking Aspen (#41) indicates that the plant is a tree (T), that the leaves are alternate (A) on the twig, that the leaves are simple (S) and not compound, that the shape is more or less cordate (C), and that the leaf margin is toothed or serrate (S). Poison Ivy (#63), a vine with alternate trifoliate leaves that usually have entire ovate leaflets, would have the code V-A/TOE. The use of this code can help you to learn and visualize leaf terms, to be more critical in your observations, and to be more accurate in your identifications.

S

T [JDP]

1. Fraser Fir
Abies fraseri

Both the Fraser Fir of higher elevations in the southern Appalachians and the very similar Balsam Fir (*Abies balsamea*) of New England and Canada have dark green, flattened, linear leaves or needles 1–2 cm (ca. .75") long with 2 white lines beneath. These striking trees may be 12–20 m (40–60') tall. The upright cones are 5–8 cm (2–3") long, and the cone scales fall with the seed.

The resin collected from blisters in the bark is known as "Canada Balsam" and was used by the Indians to caulk their birch bark canoes; later it was used as a glue for glass. The fragrant needles do not drop off as the tree dries, making firs excellent Christmas trees.

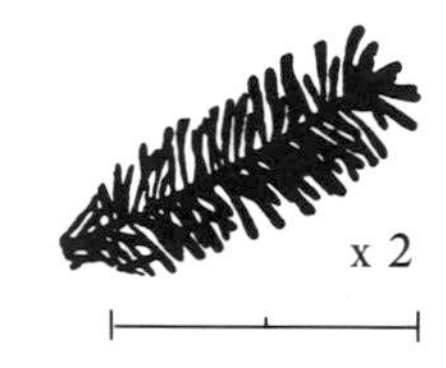

T-A/SLE Pine Family (*Pinaceae*)

2. Hemlock
Tsuga canadensis

The flattened needles of Hemlock, about 1.5 cm (ca. .5") long, also have 2 white lines beneath but leave a small rough stump on the twig when they fall. The small, pendant cones are 1.5–2.5 cm (.5–1") long, and usually occur toward the end of the somewhat flattened branches. Hemlocks may grow to 35 m (110') or more in height and up to 1 m (3') or more in diameter in rich, moist mountain coves.

The tree provides lumber and pulp for paper, and is also a valuable tree for landscape use. Hemlock bark was once an important source of tannin.

Hemlock is the state tree of Pennsylvania.

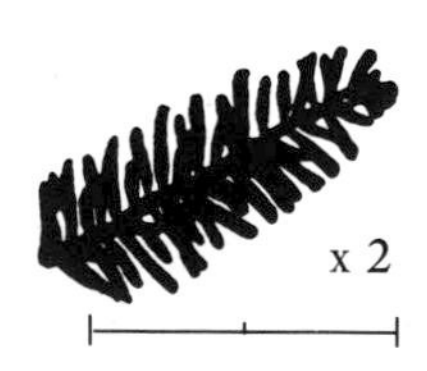

T-A/SLE Pine Family (*Pinaceae*)

1 a
2 a
1 b
2 b

3. Red Spruce
Picea rubens

The yellow-green, sharp, stiff, 4-angled needles of Red Spruce are 1 cm (ca. .5") or more long. The trees grow very slowly but can be 25–30 m (75–90') tall. They are frequent in pure stands or mixed with hardwoods in the White, Green, and Adirondack Mountains, less common in the Poconos, rare in the Smokies.

The Indians and early settlers had many uses for all parts and products of both the Red Spruce and the closely related Black Spruce (*Picea mariana*). Once in demand for ship masts, the trees are now used for pulpwood, lumber, and Christmas trees. The resonant wood is used in the production of violins and dulcimers.

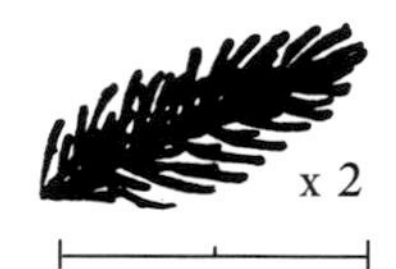

T-A/SSE Pine Family (*Pinaceae*)

4. Arbor Vitae
Thuja occidentalis

The green to yellow-green needles of White Cedar, as this plant is also called, are small, flat, scale-like, and closely appressed to the numerous branchlets that form soft, flattened sprays at the ends of the widely spreading primary branches. The cones are brown and woody; the usually abundant seeds are dispersed by the wind. A medium-sized tree, to 20 m (60'), Arbor Vitae grows in cool northern or mountain swamps and bogs, especially in the White, Green, and Adirondack Mountains.

It takes Arbor Vitae a century to grow large enough for lumber, and few such trees remain. Many horticultural forms of this tree exist, however, and it is widely used in landscaping.

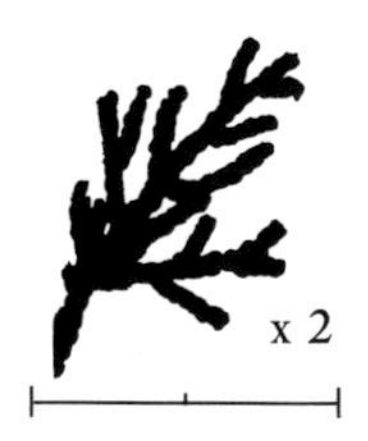

T-O/SSE Cypress Family (*Cupressaceae*)

3

4 [PSW]

5. Creeping Cedar
Juniperus horizontalis

The scale-like leaves or needles of this interesting prostrate relative of the common Red Cedar are pressed against the slender twigs. The prostrate branches of this evergreen shrub may get several meters long, and a single plant may form a conspicuous evergreen circle up to 5 m (15') or more in diameter on rocky or sandy soils in northern New England.

Creeping Cedar is easily propagated by cuttings, grows well in a number of different soils, and is thus of horticultural value as an evergreen ground cover.

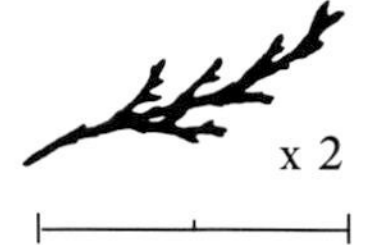

S-O/SSE Cypress Family (*Cupressaceae*)

6. Red Cedar
Juniperus virginiana

On young branches the scale-like leaves or needles of Red Cedar may be .5 cm (.25") long and spreading, but on older growth they are shorter and appressed to the twigs. This small to medium-sized tree may grow to 20 m (60') or more on limestone soils in woodlands and old fields throughout much of the eastern United States. In the spring the male trees turn yellow-brown as their thousands of small, papery pollen cones mature.

The blue-gray "berries" (#120) on the female trees of Red Cedar are actually fleshy cones that are eaten (and dispersed) by birds. The colorful, aromatic heartwood, which resists rot and repels some insects, is used for fence posts, outdoor furniture, shingles, and cedar chests. The trees grow rapidly and are often used in landscaping, as screening, or for windbreaks.

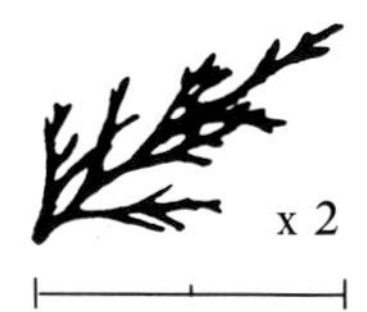

T-O/SSE Cypress Family (*Cupressaceae*)

5
6 a
6 b

7 a

7 b [DHR]

7. White Pine
Pinus strobus

The 5 slender, bluish-green needles, 7–14 cm (3–5") long, in each bundle and the elongate leathery cones assure positive identification of this handsome and widespread tree. White Pine grows rapidly on many different soil types. It may reach 30 m (90') or more in only 30 years and up to 60 m (180') when mature.

The tree is often used in landscaping and for Christmas trees, and its soft lumber is used for paneling, shelving, and molding. In Colonial times huge White Pines in the virgin forests were claimed "by the Crown" for masts for England's ships.

White Pine is the state tree of Maine.

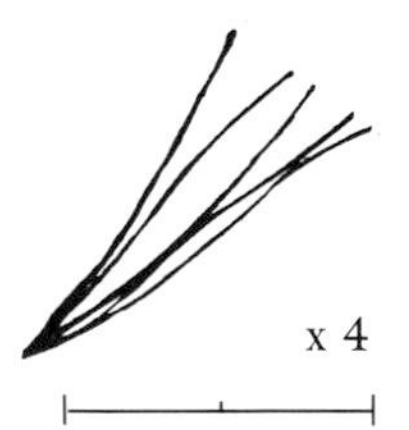

T-A/SSE Pine Family (*Pinaceae*)

7 c

8. Jack Pine
Pinus banksiana

The 2 flat, slightly twisted needles in each bundle are 2–4 cm (.75–1.5") long. The slender, spineless, yellowish brown cones are about the same length as the needles; they usually remain unopened on the tree for many years but open after the heat of a forest fire. Sometimes called Scrub Pine, the tree is small to medium in size and grows on rocky or sandy soils in the far north. It resembles the Virginia Pine, which has a more southern distribution.

Jack Pine is a pioneer species after fires or clearing. It has taken over many areas where the Paper Birch and Quaking Aspen have been logged out.

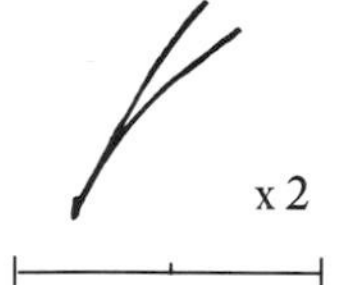

P-A/SSE Pine Family (*Pinaceae*)

9. Virginia Pine
Pinus virginiana

The 2 twisted, dark green needles in each bundle are 4–8 cm (1.5–3") long. The small, oblong cones are 4–7 cm (1.5–2.5") long and may remain on the branches for several years. The tree is often rather small, but older trees may reach 30 m (90'). A common pioneer tree on poor soil of old fields at elevations below 1,000 m (3,000'), it is vegetatively very similar to the more northern Jack Pine (*Pinus banksiana*). Both species are sometimes called Scrub Pine.

The wood is hard and often knotty, but can be used for pulp, fuel, and lumber.

T-A/SSE Pine Family (*Pinaceae*)

8

9

10. Rosebay

Rhododendron maximum

The dark, leathery evergreen leaves of this shrub or small tree are oblong and 10–20 cm (4–8") long; some individual leaves may remain on the stem for 5 years or more. In moist woods, especially in the mountains, Rosebay, also known by its scientific name of Rhododendron, may form dense, almost impenetrable colonies, and the plants may reach a height of 10 m (30') or more.

Because of its attractive evergreen leaves and its showy clusters of large flowers in the spring, Rhododendron is often used in landscaping. The wood is hard and makes good fuel; the burls were once used to make pipe bowls.

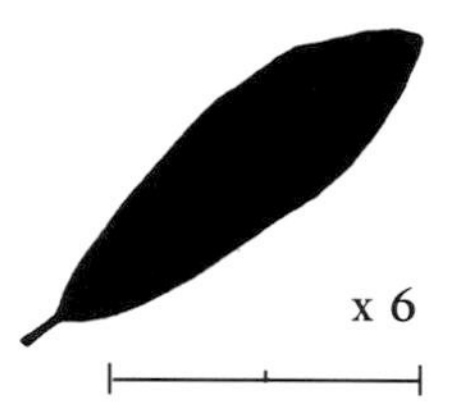

S-A/SEE Heath Family (*Ericaceae*)

11. Mountain Laurel

Kalmia latifolia

The glossy, elliptic, evergreen leaves of this shrub or small tree are 5–12 cm (2–4.5") long and turn yellow as they individually age and drop off. The plants become thick and shrubby and are very attractive on road banks and in clearings, where they get ample light. In the shade of mature deciduous forests the plants are elongate, with few branches and only a few leaves near the top.

There are a number of horticultural varieties of Mountain Laurel, or Ivy, as some mountain people call it. At one time the leafy twigs were used to make evergreen garlands for Christmas decoration.

S-A/SEE Heath Family (*Ericaceae*)

10

11

12. American Holly
Ilex opaca

The spiny, leathery evergreen leaves, 5–10 cm (2–4") long, the smooth gray bark, and the red berries on the female trees make this traditional Christmas evergreen one of the easiest trees in the woodlands to identify. Holly, a broad-leaf evergreen, grows in low deciduous woodlands primarily in the coastal plain and piedmont.

Many horticultural varieties of Holly are used in landscaping, the red berries are eaten by birds, and the ivory-white heartwood is used in furniture and for carving.

American Holly is the state tree of Delaware.

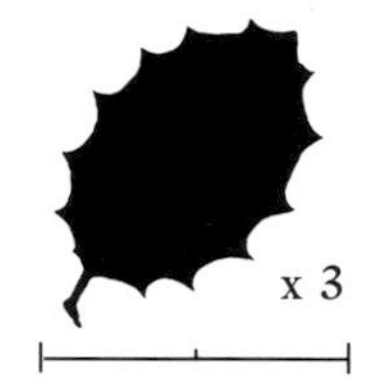

T-A/SEE Holly Family (*Aquifoliaceae*)

13. Southern Magnolia
Magnolia grandiflora

The large, glossy, dark green leaves of Southern Magnolia, up to 20 cm (8") long, make it our most spectacular broad-leaf evergreen. The bright red seeds that dangle from the cone-like fruits until eaten by birds add further color to these large trees, which are native to the southeastern coastal plain but are widely planted as ornamentals.

At one time these trees may have been plentiful enough to have value as a timber tree, but today their large, showy, fragrant flowers and their glossy evergreen foliage make them most important as a landscape plant.

Southern Magnolia is the state tree of Mississippi.

T-A/SEE Magnolia Family (*Magnoliaceae*)

12

13 [WSJ]

14. Christmas Fern

Polystichum acrostichoides

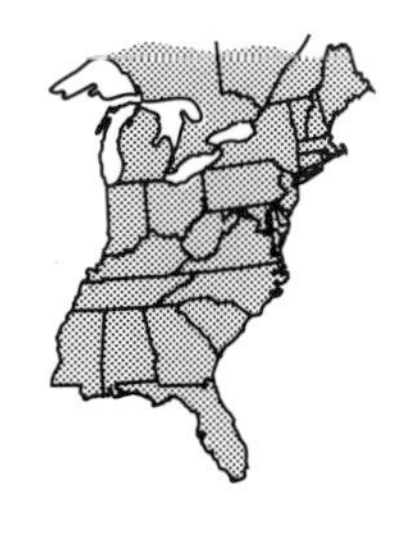

This rhizomatous, more or less evergreen fern occurs in large clumps in low, open woodlands throughout much of eastern North America. Its erect, pinnate green leaves last one year, from spring to spring, and new leaves form as the old ones die.

Christmas Fern is a favorite plant for woodland and wildflower gardens, and its firm leaves are sometimes used locally for decoration.

H-A/PNS Polypody Family (*Polypodiaceae*)

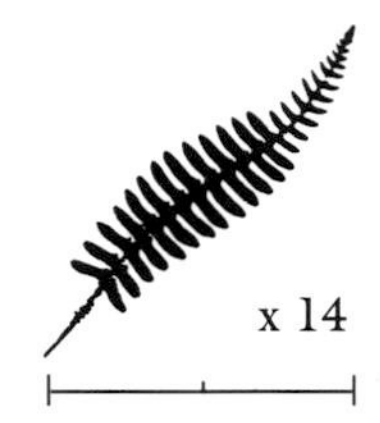

15. Rock Cap Fern

Polypodium virginianum

The evergreen pinnate leaves of this small fern are 10–20 cm (4–8") long and, though less than half the size, are similar in general outline to those of Christmas Fern. Polypody generally grows on shaded rocks and slopes or on old fallen logs. More rarely it may grow on the bark of living oaks or other hardwoods, which is the usual habitat for the closely related Resurrection Fern (*Polypodium polypodioides*).

Rock Cap Ferns, found throughout the southern Appalachians, were used by the Indians to make a medicinal tea. Today they are often used in woodland gardens.

H-A/PLE Polypody Family (*Polypodiaceae*)

14

15

16. Bracken Fern

Pteridium aquilinum

The large, triangular, twice pinnate yellow leaves of this rhizomatous weedy fern may be 1 m (3') or more tall and are produced a few inches apart near the tip of the rhizome. Different varieties of Bracken Fern occur over much of North America in open and second-growth woodlands, along roadsides, and in clearings and pastures.

Although the plant is poisonous if eaten, in earlier times the dried rhizome was used medicinally as a cure for worms.

H-A/BLE Polypody Family (*Polypodiaceae*)

17. Cinnamon Fern

Osmunda cinnamomea

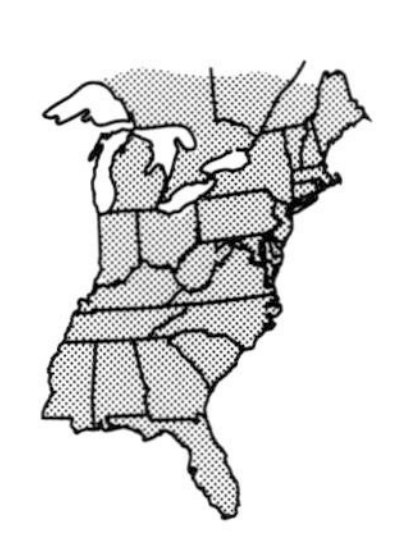

The compact yellow-brown clusters of fall fronds of Cinnamon Fern may be 1 m (3') or more tall, and the large, usually distinct clumps of this rhizomatous perennial of bogs and wet meadows may be 1 m or more in diameter. As the yellow leaves die down, the plants go dormant until the following spring, when new foliage leaves and the characteristic cinnamon-colored spore leaves are formed.

Often used as an accent plant, this attractive fern also makes a good component of a woodland or wildflower garden.

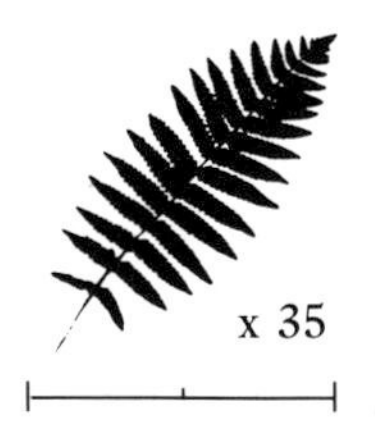

H-A/PES Royal Fern Family (*Osmundaceae*)

16 [PSW]

17 a [PSW]

17 b [RKG]

18. Cucumber Magnolia

Magnolia acuminata

The large, evenly tapered, elliptic leaves of this magnolia are 10–30 cm (4–12") long and may be reddish brown or golden yellow. The small knobby fruits, seldom more than 5–7 cm (ca. 2.5") long, resemble a cucumber only slightly when green and even less when the bright red seeds, which are eaten by birds, begin to show.

If cut off near the ground, these small to medium-sized trees send up many shoots that form a compact bush-like clump as they grow. Cucumber Magnolia is found in rich woodlands in the southern Appalachians and their adjacent foothills from the Cumberlands to the Smokies.

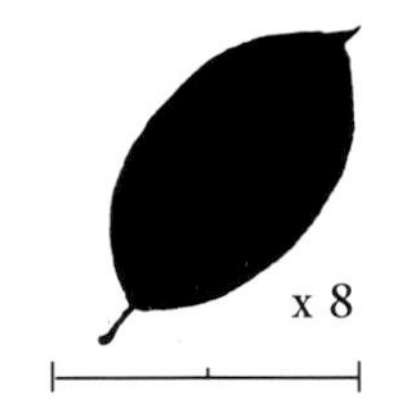

T-A/SEE Magnolia Family (*Magnoliaceae*)

19. Pawpaw

Asimina triloba

The large, obovate, yellow leaves of Pawpaw, 15–25 cm (6–10") long, are rusty pubescent beneath and have an unpleasant odor when crushed. These shrubs or small trees may be up to 10 m (30') tall and often form large clumps from root sprouts. They are frequently found in river bottom and floodplain woodlands of the southern Appalachians and adjacent areas.

The fleshy brown or purplish fruits, 8–13 cm (3–5") long, have edible yellow pulp. When the plants were more common, their fruits were gathered for food; today they are primarily eaten by woodland animals.

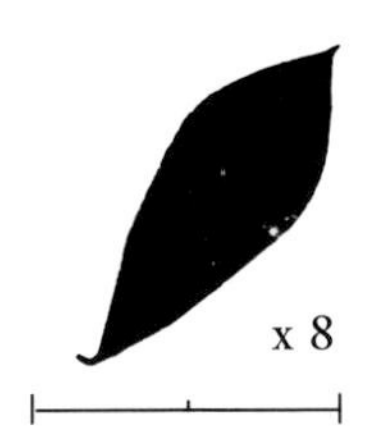

T-A/SBE Custard Apple Family (*Annonaceae*)

18 a

18 b

19

20. Spicebush
Lindera benzoin

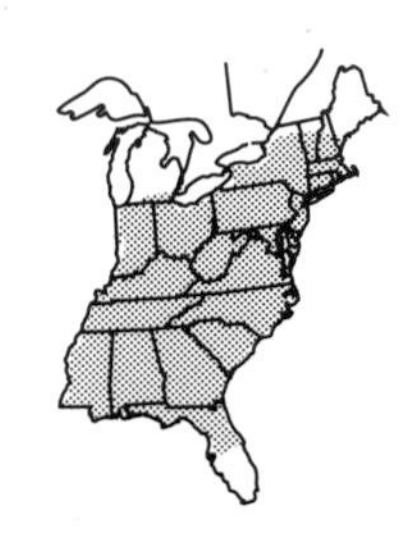

The bright yellow autumn foliage, the shiny red berries, and the spicy fragrance of the fruit, leaves, and twigs of this attractive shrub make it easy to spot and identify in floodplain or stream margin woodlands. The shrub is 1–2 m (3–6") or more tall with thin, entire, lanceolate or slightly obovate leaves 5–12 cm (2–5") long.

Because of its fragrance, early spring bloom, and colorful berries, Spicebush is sometimes used in landscaping.

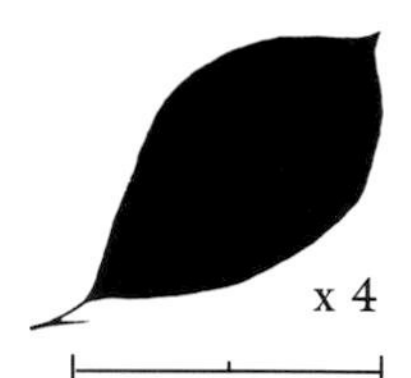

S-A/SEE Laurel Family (*Lauraceae*)

21. Redbud
Cercis canadensis

The yellow or greenish yellow, heart-shaped leaves of Redbud are 5–10 cm (2–4") long; the slender brown fruits, or pods, are 4–10 cm (ca. 2–4") long and are usually borne in clusters along the branches.

This small tree is somewhat weedy, especially in open, limestone soils, and often occurs in large populations. In the spring its masses of dark pink flowers add color to cedar glades and hardwood forests. Because of this spring color and its small size, Redbud is frequently used in landscaping.

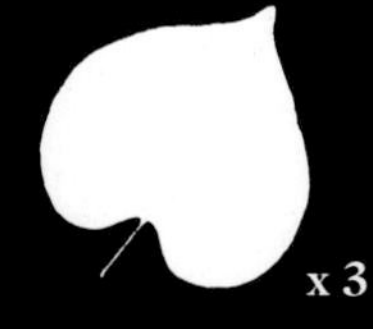

T-A/SCE Bean Family (*Fabaceae*)

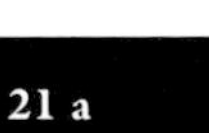

20

21 a

21 b

22. Pipe Vine
Aristolochia macrophylla

The large, heart-shaped, yellow or tawny leaves of Pipe Vine, or Dutchman's Pipe (as it is also called in reference to the shape of its distinctive brown flowers), are 8–24 cm (3–9") long or wide. They are alternate but may appear to be opposite if crowded. Since these supple vines reach the tops of trees in rich woods and along stream banks, the leaves tend to be seen only when the vines occur along the open forest edge at road cuts and other clearings.

People sometimes include this native vine in their gardens, or grow the related tropical Calico Dutchman's Pipe as a houseplant.

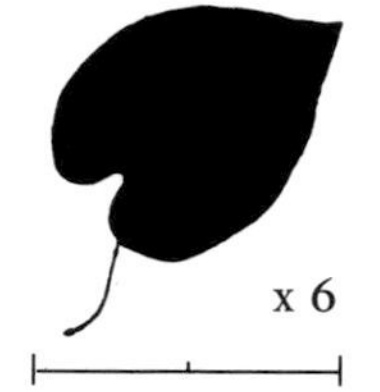

V-A/SCE Birthwort Family (*Aristolochiaceae*)

23. Fringe Tree
Chionanthus virginicus

The opposite, entire, yellowish leaves of this shrub or small tree vary from widely elliptic to obovate and are up to 20 cm (8") long. Fringe Tree grows in dry deciduous woods, on rock outcrops, and on lowland savannas throughout much of the southern Appalachians.

The showy cluster of white flowers in the spring and the dark blue fruits in the fall account for the frequent use of Fringe Tree as an ornamental.

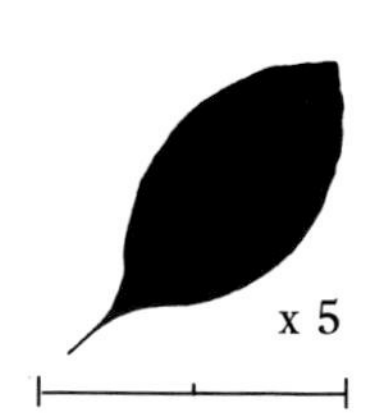

S-O/SEE Olive Family (*Oleaceae*)

22

23

24. Sweet Shrub
Calycanthus floridus

The bright yellow leaves of this aromatic colonial shrub, also known as Carolina Allspice, are opposite, ovate to lanceolate, and 5–15 cm (2–6") long, and have smooth margins. Sweet Shrub grows in hardwood forests and clearings, generally below 1,300 m (3,900') in the southern Appalachian area. The brilliant foliage is especially noticeable along mountain roadsides, where the colonial plants often form large clumps.

Because of their maroon spring flowers (which look like small, brown magnolia flowers) and the spicy fragrance of their flowers, stems, and leaves, these attractive plants are often planted in gardens. They are easy to propagate from seed (if the mice leave any!) and cuttings.

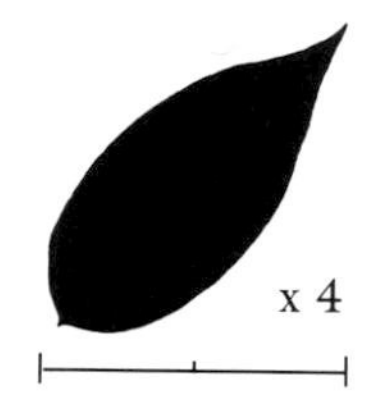

S-O/SOE Calycanthus Family (*Calycanthaceae*)

25. Larch
Larix laricina

Also called Tamarack, these small native trees of the far north are closely related to the pines. They are "deciduous evergreens," and their tufts of small slender leaves turn yellow before dropping off each fall. The closely related European Larch (*Larix decidua*) has been widely planted in Canada and New England and has become naturalized in some areas.

The European Larch is an important timber tree, and the durable wood of both species can be used for poles, posts, railroad ties, and other ground contact applications. Both species are also used, primarily in their range, as ornamentals.

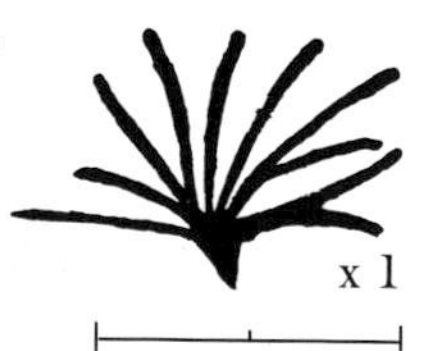

T-A/SLE Pine Family (*Pinaceae*)

24

25 a

25 b

26. Black Cherry
Prunus serotina

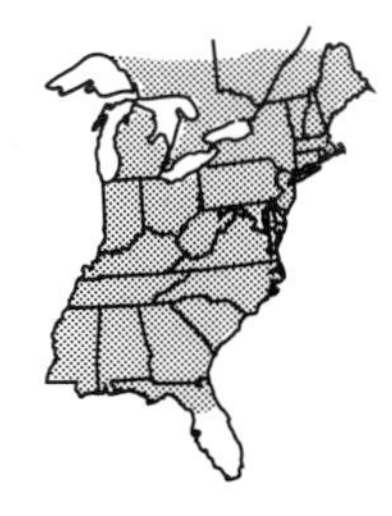

The greenish yellow to gold lanceolate leaves of Black Cherry are 5–10 cm (2–4") long and finely serrate. They have a bitter taste and are poisonous to livestock if eaten.

The cherries are eaten by birds and other wildlife and are gathered by people to make jelly or wine. Black Cherry syrup, made from the bark, has been a standard cough medicine for a century. The dark, fine-grained wood ranks with walnut for use in fine furniture, paneling, and a number of special uses.

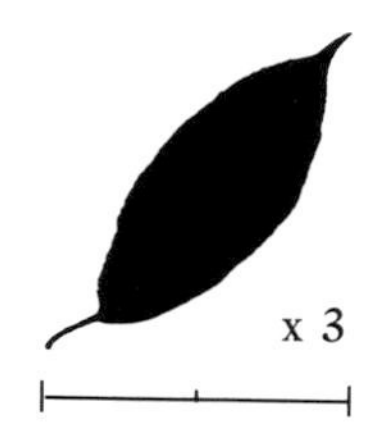

T-A/SNS Rose Family (*Rosaceae*)

27. Fire Cherry
Prunus pensylvanica

The maroon to red or orange lanceolate leaves of Pin Cherry, as it is also known, are 5–15 cm (2– 6") long, have a bitter taste, and are borne on bright red twigs. These fast-growing shrubs or small trees quickly invade clearings, especially clearings caused by fire, and can form large stands. Primarily a northern species, Fire Cherry follows the crest of the Appalachians through the Smokies to northern Georgia.

The cherries can be made into jelly, but most are consumed by birds and other wildlife. The leaves are poisonous to livestock if eaten.

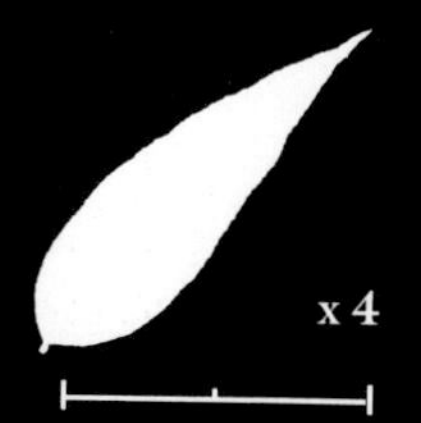

T-A/SNS Rose Family (*Rosaceae*)

26

27a

27b [BGH]

28. Witch Hazel

Hamamelis virginiana

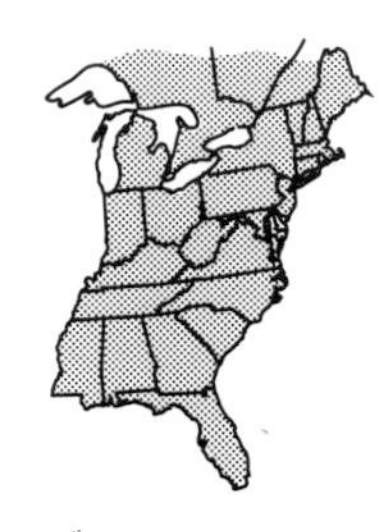

The clear yellow to golden yellow leaves of Witch Hazel are ovate to obovate, are 5–15 cm (2–6") long, and have shallow rounded lobes or teeth and an oblique base. These widespread shrubs or small trees normally bloom in the fall. The bright yellow flowers are often on the branches along with the leaves and last year's hard brown capsules, which explode when dry and throw the black seeds up to 10 m (30').

An alcohol extract of the bark and young twigs of Witch Hazel is an astringent lotion with medical and cosmetic uses.

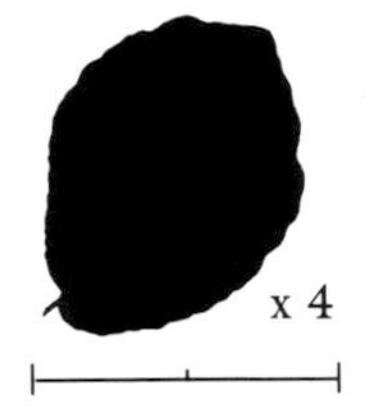

S-A/SOL Witch Hazel Family (*Hamamelidaceae*)

29. Bittersweet

Celastrus orbiculatus

The round, light yellow leaves of this woody vine, 4–7 cm (ca. 1.5–3") in diameter, and the small, axillary clusters of yellow fruits add color to thickets, fencerows, and woodland borders at scattered localities from Connecticut to North Carolina, wherever these introduced plants have spread from cultivation and become naturalized. The fruits of our native Climbing Bittersweet (*Celastrus scandens*) are in a larger, and terminal, cluster (#109).

The colorful yellow fruits of both of these vines split open and expose bright red seeds, making them attractive in the garden or in dried arrangements.

V-A/SRS Staff-tree Family (*Celastraceae*)

28

29

30. Forest Grape
Vitis riparia

The heart-shaped, coarsely serrate yellow leaves of Forest Grape, or Riverbank Grape as it may also be called, are 10–20 cm (4–8") long.

These high-climbing woody vines are found in low riverbank woods at scattered localities through much of the Appalachian area, but they are rare in the southern part of their range.

The dark, thin-skinned grapes, borne in small bunches, are eaten by most birds and animals, and by man if any are left!

V-A/SCS Grape Family (*Vitaceae*)

31. Muscadine Grape
Vitis rotundifolia

The clear yellow, heart-shaped leaves of Musca-dine Grape often form vertical streaks of color along forest margins, where the high-climbing vines may reach the top of the tallest trees.

The domesticated varieties of this dark, thick-skinned native grape include the golden brown Scuppernong, which is widely planted in the south and makes excellent wine. The large grapes are not in bunches but are solitary or in small clusters of two or three.

V-A/SCS Grape Family (*Vitaceae*)

30

31

32

32. Red Mulberry

Morus rubra

The yellow leaves of Red Mulberry are 5–20 cm (2–8") long, about the same size as those of White Mulberry, but not as glossy and more sharply pointed. Although this small tree is native to our hardwood forests, it has been spread widely by birds to fencerows, clearings, and roadsides over much of the eastern United States.

The sweet, juicy fruits are eaten by man, beasts, and birds, and the wood can be used for fence posts and furniture. The inner bark of young shoots was used by some Indians to make a crude cloth.

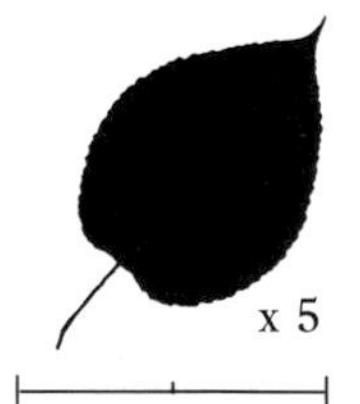

T-A/SES Mulberry Family (*Moraceae*)

33. White Mulberry
Morus alba

The bright, glossy yellow of the autumn leaves of the Silk Mulberry, as the tree is also called, provides a colorful accent in many vacant city lots, industrial areas, and open spaces. The serrate leaves are about 5–20 cm (2–8") long and may be divided into 3 lobes. This small tree, introduced in Colonial times to supply food for silkworms when silk was thought of as a possible new industry, has become naturalized and rather weedy. Its seeds have been widely spread by birds, which eat its sweet, juicy berries.

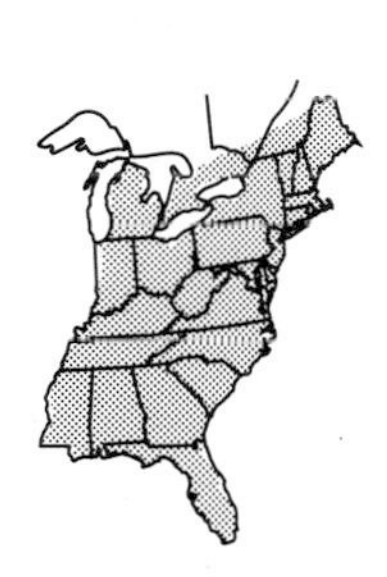

Because of its tolerance of city environments and because its berries are enjoyed by people as well as birds, the tree is sometimes used in landscaping or for windbreaks.

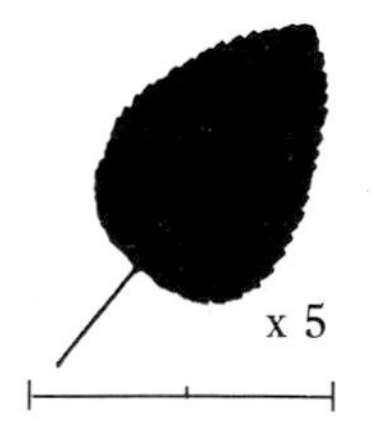

T-A/SOS Mulberry Family (*Moraceae*)

34 a

34. Beech

Fagus grandifolia

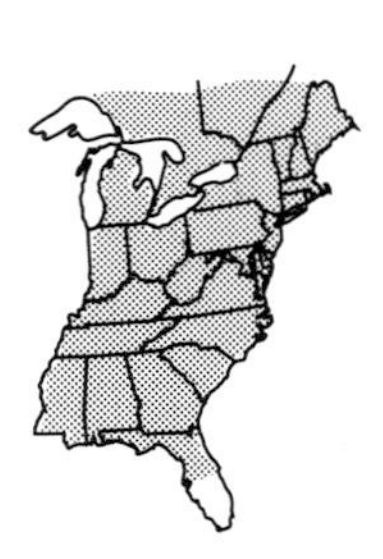

The rich yellow to bronze elliptic leaves of Beech are 5–12 cm (2–4.5") long, have straight, evenly spaced secondary veins (which color last), and often remain on the tree until well into winter. In winter this tree is easily identified by its brown leaves, slender apical buds, and smooth gray bark.

In addition to the important contribution beechnuts make to the mast crop for wildlife from birds to bears, the trees have considerable horticultural value. The hard, white, fine-grained wood has many special commercial uses, and the limbs and branches make good firewood.

T-A/SES Beech Family (*Fagaceae*)

34 b [BGH]

34 c [BGH]

34 d

The Birches

For many people the clear yellow leaves of the birches and aspens (which belong to the Willow family) are the essence of fall color. This is especially true of Paper Birch and Quaking Aspen, both of which extend across Canada and the northern United States from Labrador to Alaska.

Three other birches — Sweet Birch, Yellow Birch, and Gray Birch — also add to the fall color of the northern Appalachians. River Birch, essentially a southeastern species, is far less colorful.

All birches are wind-pollinated and produce small pendulous clusters (called "catkins") of male and of female flowers each spring before the leaves are out on the trees. The fruits are small winged nutlets that are dispersed by the wind — if not eaten by birds or squirrels.

The wood of the different species varies in structure, hardness, and use. Yellow Birch is an excellent hardwood; the softer wood of Paper Birch and River Birch is used for pulpwood.

35. Paper Birch
Betula papyrifera

The clear yellow, ovate, long-pointed leaves of Paper Birch or White Birch are 5–10 cm (2–5") long and are coarsely doubly serrate. The bright white bark of the younger trunks is what makes this distinctive tree such a focal point in the fall landscape and one of our most quickly recognized trees. Because birch bark canoes were used by the Indians and by colonial explorers and trappers, Paper Birch is also known as Canoe Birch.

These colorful trees are widely used in cooler climates as a landscape tree, but are quite rare, even in the mountains, south of their northern Appalachian range.

White Birch is the state tree of New Hampshire.

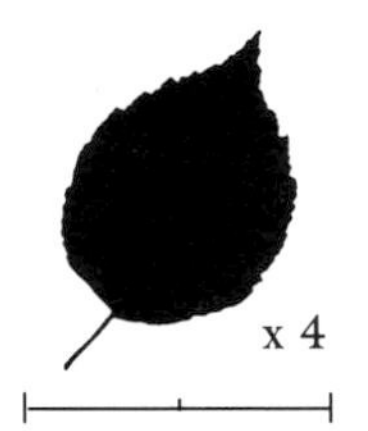

T-A/SOS Birch Family (*Betulaceae*)

36. Gray Birch
Betula populifolia

The pale yellow or orange-yellow leaves of Gray Birch are strongly triangular and 5–10 cm (2–4") long, with a long slender tip and doubly serrate margins. The bark of these small trees may be white to gray or almost green, but it is tight and does not peel like that of Paper Birch and River Birch. These pioneer trees of the northern Appalachian forests establish readily on road banks, in burned areas, and in old fields.

Gray Birch is an attractive and commonly used landscape tree. The wood is used for fuel and the manufacture of specialty items.

T-A/SCS Birch Family (*Betulaceae*)

35 a

35 b

36 [SR]

37. Sweet Birch
Betula lenta

The yellow leaves with parallel lateral veins and the smell of wintergreen from the crushed brown twigs of this medium to large cool-climate tree identify it as Sweet Birch. It is also known as Cherry Birch because the bark resembles that of Cherry. The aromatic elliptic leaves are sharply serrate and are 2.5–10 cm (1–4") long. A northern species, it follows the mountains south to the Smokies, where it is less frequent than Yellow Birch.

Oil of wintergreen or birch oil was at one time distilled from the bark and twigs of Sweet Birch, but this oil is now produced synthetically. The sweet sap of these trees can be fermented to make birch beer.

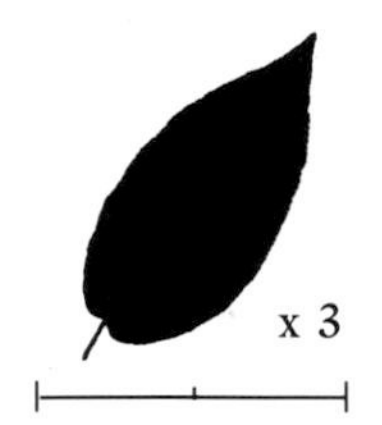

T-A/SES Birch Family (*Betulaceae*)

38. River Birch
Betula nigra

The ovate, doubly serrate leaves of River Birch, 2.5–8 cm (1–3") long, may turn yellow in the fall or they may directly turn brown and drop from the twigs. However, the thin sheets of peeling rose-brown bark on the younger stems and the thick, deeply fissured brown and gray bark of older trunks add both color and texture to autumn roadsides and landscapes at lower elevations.

These native trees, which often have 2–3 trunks, have become somewhat weedy in the southeast, but their ability to grow in a variety of soils has made them desirable as landscape plants.

T-A/SOS Birch Family (*Betulaceae*)

37 a

37 b [BGH]

38 a

38 b

39. Hop Hornbeam
Ostrya virginiana

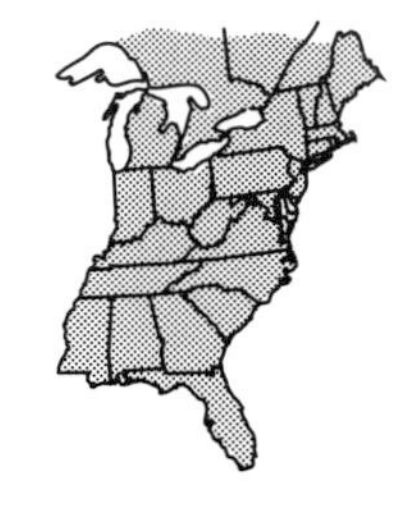

The bright yellow leaves of this small to medium-sized tree are narrowly ovate, doubly serrate, and 2.5–15 cm (1–6") long. The brown, somewhat shreddy bark and the unlobed bracts attached to the fruit separate Hop Hornbeam from Ironwood, with which it may be found.

The wood of Hop Hornbeam is very hard and has limited commercial use. The small, hard fruits are eaten by wildlife.

T-A/SOS Birch Family (*Betulaceae*)

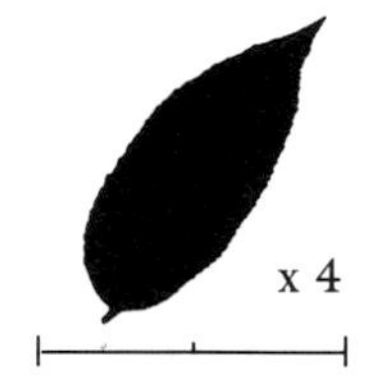

40. Ironwood
Carpinus caroliniana

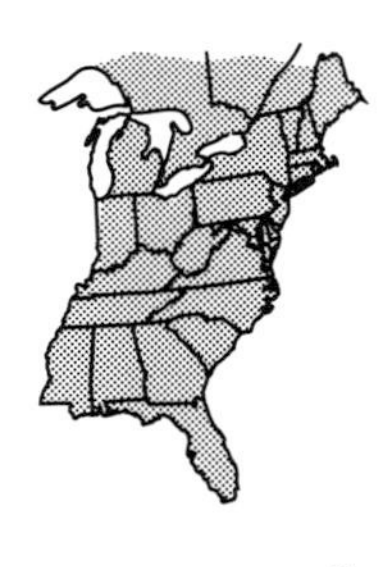

The red to orange elliptic leaves of this small tree are usually 2.5–10 cm (1–4") long and are doubly serrate. The small fruits are attached to a 3-lobed, leaf-like bract (#132) that aids the wind dispersal of the seeds. The uneven growth in stem circumference produces muscle-like ridges under the smooth gray bark, thus the name Musclewood. This understory tree is found throughout eastern North America on rich soils along streams and generally below 1,000 m (3,000').

The wood, which is very hard and smells somewhat like black pepper when dry, is used for tool handles and fuel.

T-A/SES Birch Family (*Betulaceae*)

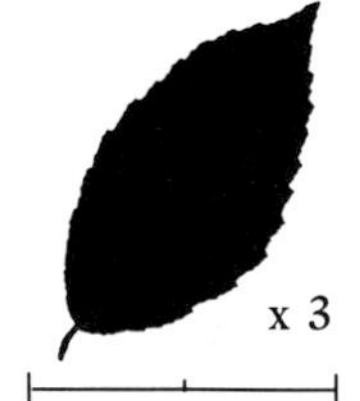

39 [RVP]

40 a

40 b

41. Quaking Aspen

Populus tremuloides

The bright yellow "quaking" leaves of this widespread tree bring brilliant fall color not only to large areas of northeastern North America, but to much of Canada and Alaska, the mountains of many western states, and even a bit of northern Mexico. The cordate, finely serrate leaves are about 4–8 cm (1.5–3") long or broad; the slender, flattened leaf stalks permit the leaf blades to tremble with the slightest breeze.

These trees are an important scenic aspect of many areas and are used in landscaping. The soft wood is used for pulp, boxes, tubs, and paneling.

T-A/SCS Willow Family (*Salicaceae*)

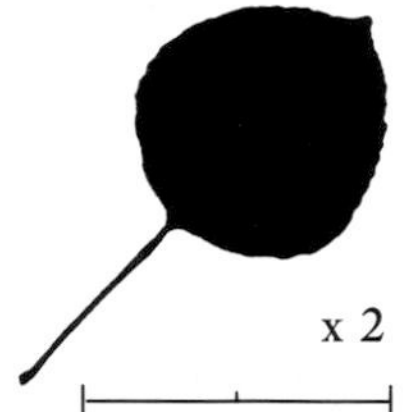

x 2

42. Big-tooth Aspen

Populus grandidentata

The large yellow leaves of this small northern tree are sometimes tinged with red. The ovate leaves are 6–10 cm (2.5–4") long and the blade margin has 6–8 prominent teeth on each side, clearly distinguishing this species from Quaking Aspen. Both species pioneer areas cleared by fire or logging, and they often occur together over the more restricted range of Big-tooth Aspen.

The wood of this short-lived tree is used for pulp, fuel, boxes, and excelsior.

T-A/SOS Willow Family (*Salicaceae*)

41 [SR]

42 [LO]

43 a [BGH]

43. Tulip Poplar
Liriodendron tulipifera

The tall symmetrical cones of light golden yellow foliage that punctuate the reds, browns, and greens of many fall hillsides of the southern Appalachians are Tulip Poplars. This species has one of the most distinctive leaves of any of our trees, 8–20 cm (3–8") long or wide and broadly notched at the apex.

The light, winged fruits shed from the woody, cup-shaped "cones" (#133) are spread by the wind and are thus often the first seeds to reach any site that has been cleared. When the Chestnuts died from the blight half a century ago, Tulip Poplars "moved in" and now form large, even-aged stands where Chestnuts once grew. The tall, straight trees are very valuable for plywood, furniture, and lumber, and the large flowers are a prime nectar source for the honeybee.

Tulip Poplar is the state tree of Tennessee and Indiana.

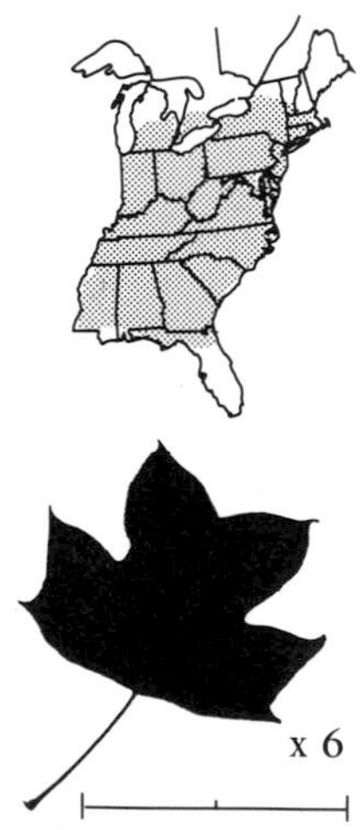

T-A/SRL Magnolia Family (*Magnoliaceae*)

43 b

43 c

43 d

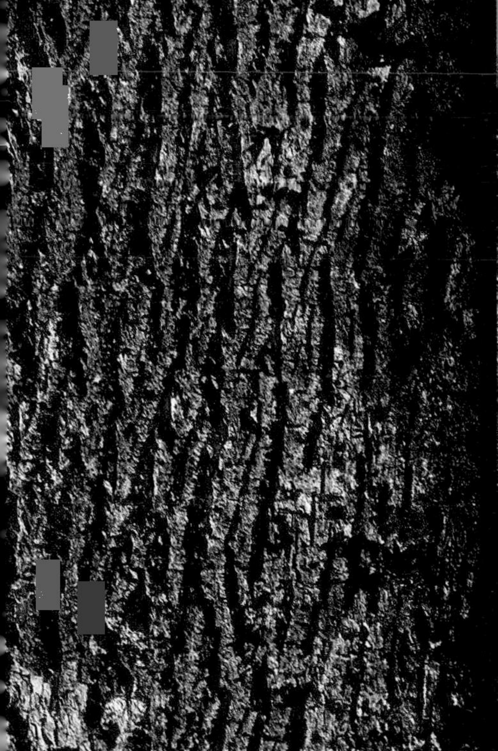

43 e [BGH]

44 a

44 b

44. American Elm

Ulmus americana

The golden yellow, ovate, serrate leaves of American or White Elm may be smooth or rough on the upper surface. They are 8–15 cm (3–6") long, usually oblique (uneven) at the base, and more gradually tapered to the pointed apex than those of Slippery Elm. This is shown in figure 44 b, where the top 2 leaves are American Elm, the middle leaf is Slippery Elm, and the lower 2 leaves are Winged Elm.

These tall, graceful trees were once widely planted along city streets and around large buildings, but many of these trees, as well as many elms in the forest, have been killed by the Dutch Elm disease. Though widespread, American Elm is generally absent from forests above 700 m (2,000') elevation.

American Elm is the state tree of Massachusetts.

T-A/SOS Elm Family (*Ulmaceae*)

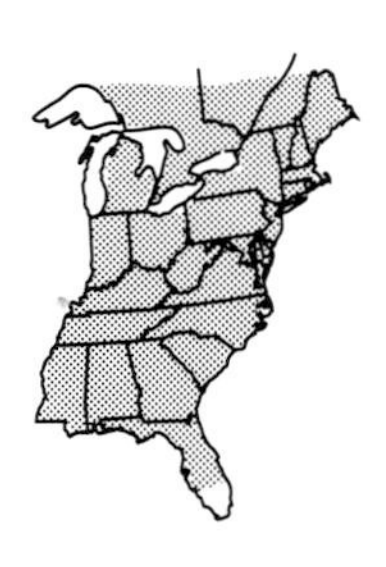

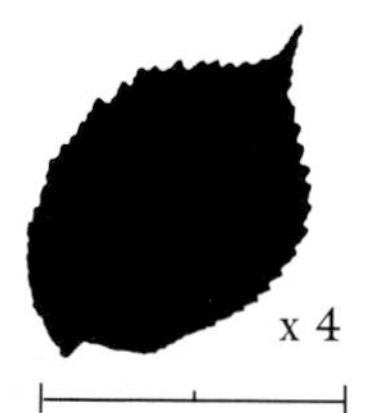

44 c

45. Slippery Elm

Ulmus rubra

The yellow to golden brown leaves of Slippery or Red Elm (see center leaf, figure 44 b) are more sharply and abruptly pointed than those of American Elm, and are very rough on the upper surface. It is the young inner bark that is slippery, not the leaves.

These small trees of our deciduous forests at lower elevations are also very susceptible to the Dutch Elm disease and are dying out in many areas. The hard wood is used for furniture, lumber, fence posts, railroad ties, and fuel.

T-A/SOS Elm Family (*Ulmaceae*)

x 4

46. Winged Elm

Ulmus alata

The rich golden yellow leaves of the Winged Elm, 4–7 cm (1.5–2.5") long, are smaller than those of the other two common elms but are equally attractive in the fall. The corky ridges or "wings" often found on young twigs account for the common name of this more southern species. Like the previous two elms, Winged Elm grows only at lower elevations.

The hard wood of this elm can be used for posts and fuel.

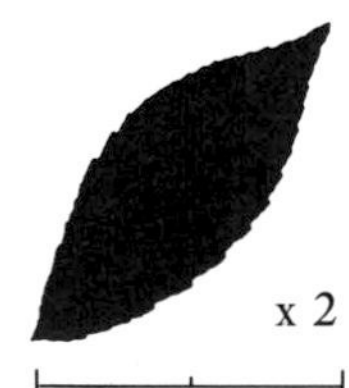

x 2

T-A/SOS Elm Family (*Ulmaceae*)

45

46 a

46 b

47. Chestnut
Castanea dentata

The long, elliptic, sharply serrate leaves of Chestnut (now found only as stump sprouts — and even these are becoming more rare) may be yellow or brown and are 10–25 cm (4–10") long.

In Colonial times these magnificant monarchs dominated our forests and supplied a bounty of food and a steady supply of durable, easy to split wood for cabins, rail fences, shingles, and fuel. Probably no other species of tree was as important to the welfare of the early settlers as Chestnut. The removal of these trees from our forests by the imported chestnut blight at the turn of the century drastically changed both the ecology and the commerce of eastern North America.

T-A/SES Beech Family (*Fagaceae*)

48. Black Walnut
Juglans nigra

The large, pinnately compound, light yellow or brown leaves of Black Walnut have a pungent odor and are 30–60 cm (12–24") long with 11–17 ovate to lanceolate, finely serrate leaflets, the terminal leaflet being small or absent. The leaves of White Walnut or Butternut (*Juglans cinerea*) are similar but usually have a terminal leaflet. Also, White Walnut's nuts are oblong, Black Walnut's round.

It takes these exceptionally valuable timber trees up to a century to grow to 30 m (90') or more in rich bottomland soil, and few such mature trees remain. The dark, hard, strong, fine-grained wood is in great demand for fine furniture, gunstocks, and paneling. The tasty nutmeats are used in candy, in baking, and for oil in flavoring. The outer husks of the round 4–6 cm (1.5–2.5") fruits (#140) furnish a brown die.

T-A/PNS Walnut Family (*Juglandaceae*)

47 a

47 b [WSJ]

48 a

48 b

The Hickories

The golden yellow of the hickories is one of the richest colors in the fall forest. Their leaves are pinnately compound, with 5–9 generally lanceolate or ovate leaflets per leaf. Two of our common hickories, Pignut and Shagbark, usually have only 5 leaflets per leaf (see page 78); two others, Bitternut (left in above figure) and Mockernut (right, above) usually have 7–9.

The often bountiful crop of nuts is an important part of the mast crop which feeds the squirrels, chipmunks, and other woodland rodents. The strong white wood of hickory is used for tool and implement handles, wagon wheel spokes, skis, furniture, paneling, lumber, and heavy timbers. The oil of the nuts has been used medicinally, and the Indians made a "milk" by pounding the sweet, oily kernels in hot water. In Colonial times a yellow dye was made from the inner bark of Shagbark Hickory. Pecan (*Carya illinoensis*), native to the Mississippi Valley, is a widely cultivated hickory of commercial value both for food and for lumber.

49. Shagbark Hickory
Carya ovata

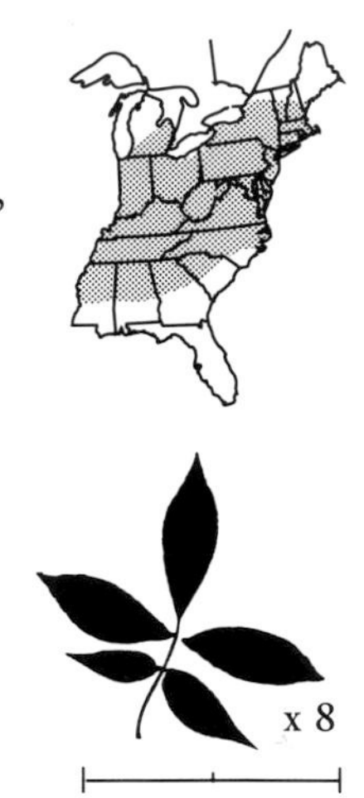

The golden yellow, pinnately compound leaves of this distinctive 5-leaflet hickory usually have 4 elliptic, finely serrate lateral leaflets and a larger, obovate terminal leaflet 8–18 cm (3–7") long. The older trees have long, shaggy strips of light gray bark. The round fruits, 3–6 cm (1.5–2.5") in diameter, have a thick husk around a slightly 4-angled nut, which is sweet and edible but hard to crack!

T-A/PES Walnut Family (*Juglandaceae*)

50. Pignut Hickory
Carya glabra

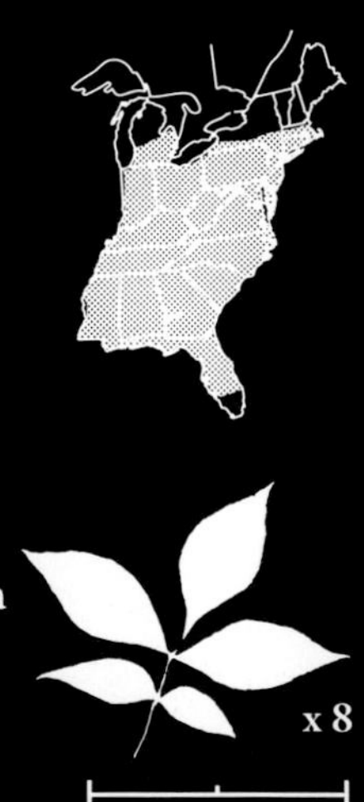

The rich yellow leaves of this common hickory of the southern Appalachian area usually have only 5 ovate-lanceolate leaflets; the terminal leaflet is the largest and is 10–18 cm (4–7") long. The bark of Pignut Hickory is light to dark gray with prominent ridges and furrows, but it does not become "shaggy." The obovoid or slightly top-shaped fruits are 2.5–5 cm (1–2") long with a relatively thin husk around a thick-shelled nut that may be sweet or bitter.

T A/PNS Walnut Family (*Juglandaceae*)

49 a

49 b

50 a

50 b

51. Mockernut Hickory
Carya tomentosa

The golden yellow, sharply aromatic leaves of Mockernut Hickory usually have 7–9 elliptic leaflets that are very hairy beneath; the three end leaflets, 10–15 cm (4–6") long, are the largest. The tight gray bark has low rounded ridges and shallow furrows. This common hickory is found in dry, upland oak-hickory forests primarily in the southern Appalachian area. The nuts (#141) are sweet and nutritious, but it takes a long time to get enough of the meats for a snack!

T-A/PES Walnut Family (*Juglandaceae*)

52. Bitternut Hickory
Carya cordiformis

As the common name indicates, this common and widespread tree of our mixed hardwood forests has nuts that are bitter and inedible. Bright yellow buds occur at the ends of the twigs, and the leaves have 7–9 leaflets that are only slightly hairy beneath.

T-A/PES Walnut Family (*Juglandaceae*)

51 a

51 b

52

53. Black Locust

Robinia pseudoacacia

The rich yellow (or pale gray-green in dry years), pinnately compound leaves of Black Locust are 20–30 cm (8–12") long and have a terminal leaflet plus 8–18 roughly paired, elliptic, entire leaflets 3–5 cm (ca. 1–2") long that often fall separately from the primary leaf stalk. Young trees have a pair of sharp thorns at the base of each leaf, but no thorns are present when these fast-growing trees reach mature size, 15–20 m (45–60') or more.

These trees are often weedy and spread rapidly to abandoned fields and roadsides. Because of their showy, fragrant flowers each spring, they are widely planted as ornamentals; they are also used for erosion control. The very hard wood was once used for the hubs of wagon wheels and for pulleys in early mills; today it is used for fence posts and for the pilings of freshwater docks.

T-A/PEE Bean Family (*Fabaceae*)

54. Rugosa Rose

Rosa rugosa

The yellow pinnate leaflets of this introduced seaside rose are 2.5 cm (1") or more long, and the colorful fruits are nearly 2.5 cm (1") in diameter. There are a dozen or more native and naturalized species of roses in eastern North America that all have dull to bright red fruits, called "hips," and may have a bit of yellow in their fall foliage. None, however, can match the color of this garden rose from Japan, which has occasionally become naturalized along the New England coast.

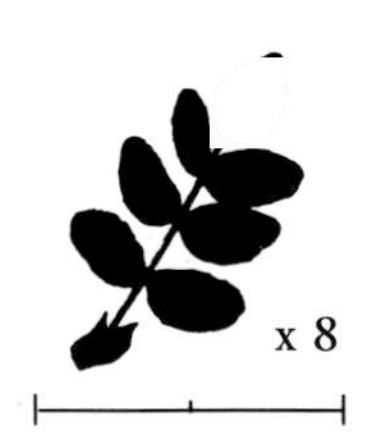

S-A/PES Rose Family (*Rosaceae*)

53

54
[WHG]

55 a [BGH]

55 b [BGH]

55. Mountain Ash
Sorbus americana

The glabrous, pinnate leaves of this native shrub or small tree may be either a dull red or a rich yellow in the fall and offer a colorful contrast to its clusters of bright red fruits, which may give color to an entire hillside after the leaves have dropped. The leaves have 13–17 elliptic, serrate or doubly serrate leaflets 5–10 cm (2–4") long. Mountain Ash is a northern plant of rocky outcrops that reaches the Smokies along the high mountains. The closely related European Mountain Ash (*Sorbus aucuparia*), an early introduction from Europe with hairy leaves, is now naturalized in the northern tip of the Appalachians.

The fruits (figure 55 c and #125) provide food for wildlife, but since they last so long into the winter they must not be very tasty!

T-A/PES Rose Family (*Rosaceae*)

55 c [BGH]

56. White Ash

Fraxinus americana

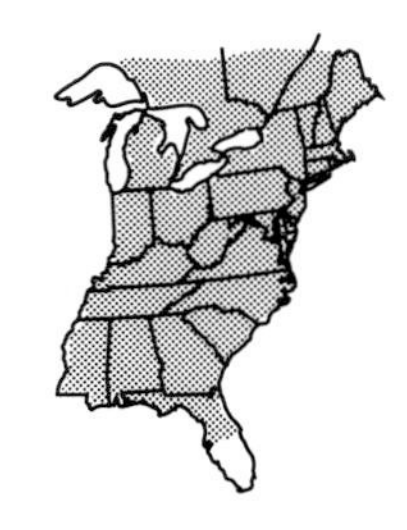

White Ash has an array of autumn colors, from yellow to red to maroon or brown, that often appear while much of the surrounding forest is still green. The usually 7 leaflets of the opposite, pinnately compound leaves are ovate to elliptic, very finely serrate, and 5–13 cm (2–5") long. These trees grow mixed with other species in moist deciduous woodlands throughout the eastern United States.

The strong, white wood of White Ash was used in the last century for spokes for wagon and buggy wheels; today it is used for baseball bats, skis, and other sporting equipment. The trees are also planted as ornamentals, as shelter belts, and for erosion control.

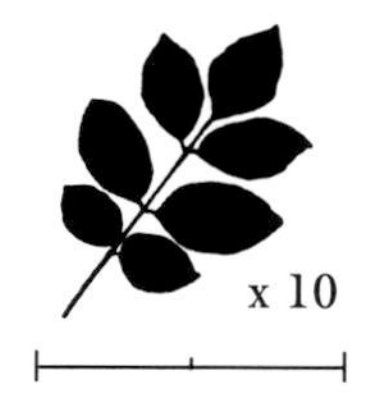

T-O/PES Olive Family (*Oleaceae*)

57. Green Ash

Fraxinus pennsylvanica

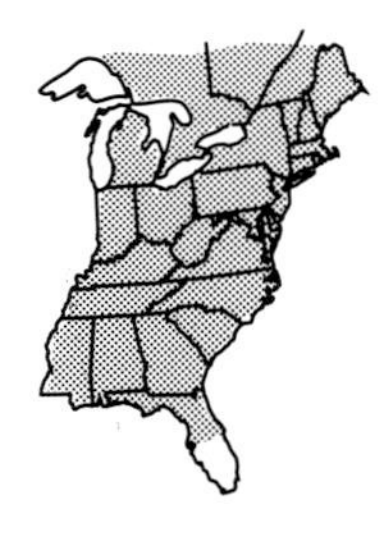

The opposite, pinnate leaves of Green Ash are uniformly pale to rich yellow over its extensive range. The 5–9 (usually 7) lanceolate to ovate leaflets are 5–13 cm (2–5") long and are very finely serrate or may be entire. The fruits are winged and wind-dispersed (#134). Green Ash grows below 1,000 m (3,000') elevation in moist woodlands and along streams throughout the eastern deciduous forests.

These large trees grow rapidly and are used to replant spoil areas after surface mining or landfill operations. They are also valuable timber trees, although the wood is not quite the quality of that of White Ash.

T-O/PEE Olive Family (*Oleaceae*)

56 a

56 b

56 c [PSW]

57

58 a

58. Sassafras
Sassafras albidum

The brilliant yellow, orange, or red leaves of this shrub or small tree are 6–15 cm (2.5–6") long and may be unlobed, have a rounded lobe on only one side, or have two side lobes. These plants are dioecious, and the female plants bear fleshy black fruits on red stalks. This is one of our most colorful fall trees along fencerows, roadsides, and woodland margins throughout most of eastern North America.

The bark of the root makes an excellent tea. The very young spring leaves are dried and powdered to thicken soups and stews (the filé gumbo of Creole cooking). The aromatic wood was once used for bedsteads in the belief that the spicy fragrance would repel insects.

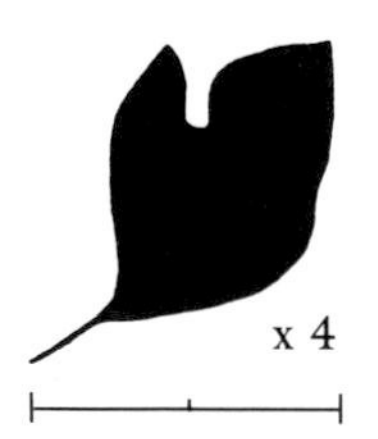
x 4

T-A/SEL Laurel Family (*Lauraceae*)

58 b

58 c

58 d [BGH]

The Maples

For many people, especially in New England, the maples typify fall color. Thanks in part to their large numbers, prominent size, and wide distribution, no other group of trees contributes as much color, brilliance, and beauty to the eastern forests. Individual trees vary in their leaf chemistry. Sugar Maple, for example, is usually yellow (top, right), but it can also be bright red or orange; Red Maple is indeed usually red (bottom, right), but it too can be shades of yellow or orange, depending on the plant's interaction with its environment. The leaves of maples are opposite and, except for Box Elder, they are simple and palmately lobed.

Maple sugar, probably the most widely known maple product, is produced commercially by boiling down the sweet, watery spring sap of Sugar Maple or the very closely related Black Maple. It takes about 32 gallons of sap to make a gallon of maple syrup.

The hard, fine-grained white wood of maple is used for furniture, flooring, paneling, musical instruments, tool handles, cutting boards, butcher blocks, wooden bowls, cooking utensils, sleds, and wagons and for many other home, farm, and industrial uses. With about 19,000,000 BTU's per cord (this is equal to 145 gallons of fuel oil), maple is also an excellent wood for fuel.

Various species with good form and color are used in landscaping or planted as shade trees along city streets. The maple leaf, as an emblem, is the central design of the Canadian flag and is also used on Canadian stamps and coins.

59 a [PSW]

59. Sugar Maple

Acer saccharum

The bright yellow, orange, or red leaves of the Sugar Maple are 10–15 cm (4–6") wide and long, have 3–5 blunt lobes, and are beautifully displayed on the compact, rounded crown of individual trees that grow in the open. In the northern part of their range, the tall trees may form nearly pure stands that can turn an entire valley or mountainside bright yellow in the fall.

The Indians showed the early settlers from Europe how to get the distinctively flavored sugar from these trees, and with some changes in technology "sugaring" remains an important part of the culture, and economics, where these trees are abundant.

Sugar Maple is the state tree of Vermont, New York, and West Virginia.

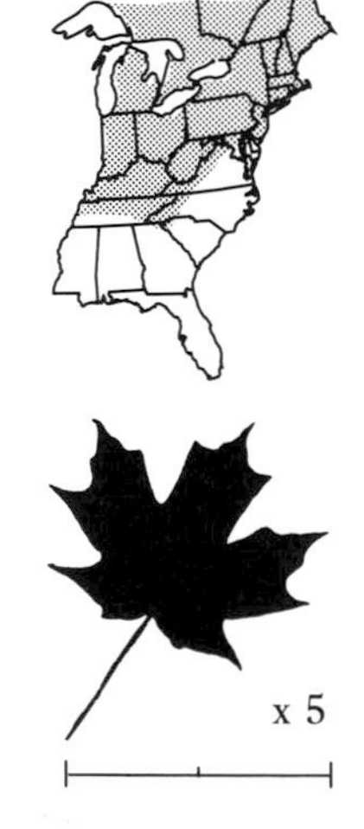

T-O/SCL Maple Family (*Aceraceae*)

59 b

59 c

59 d [SJS]

60 a [SJS]

60. Red Maple

Acer rubrum

Red Maple's brilliant red, orange, or sometimes yellow leaves are usually 3-lobed and 6–13 cm (2.5–5") long; the lobes are often pointed rather than blunt. The leaves match those of Sugar Maple for fall brilliance, but since Red Maples are generally smaller trees and tend to be scattered in the forest rather than in pure stands, they rarely dominate the fall color pattern of a given area. With a range from Newfoundland to southern Florida, Red Maples have had to adapt to more different seasonal light and temperature cycles than any other eastern tree.

The small to medium size of Red Maple makes it an ideal tree for home landscaping.

T-O/SCL Maple Family (*Aceraceae*)

60 b

61. Striped Maple
Acer pensylvanicum

The three shallow lobes at the apex of the wide, uniformly golden yellow leaves of this small understory tree give it another common name: Goosefoot Maple. The leaves are 12–18 cm (5–7") long, with doubly serrate margins. Because of their fall leaf color and the attractive green and white vertical stripes in the bark, these trees are often planted as ornamentals.

The green and white bark pattern suggests that the young stems of these trees perform photosynthesis, or food manufacture, at a low rate throughout the winter, a time when the canopy trees are bare and more light is available near the ground. If this is so, the bark would be both sweet and nutritious; and so it appears to be, since deer, rabbits, and other forest herbivores nibble it eagerly in winter.

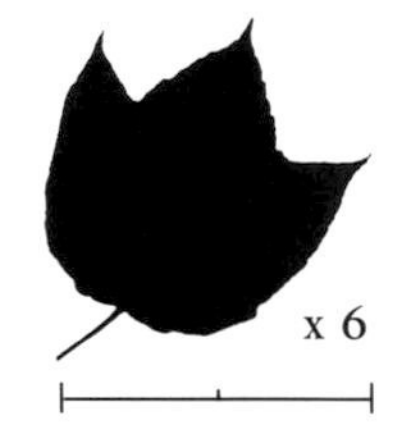

T-O/SCL Maple Family (*Aceraceae*)

62. Box Elder
Acer negundo

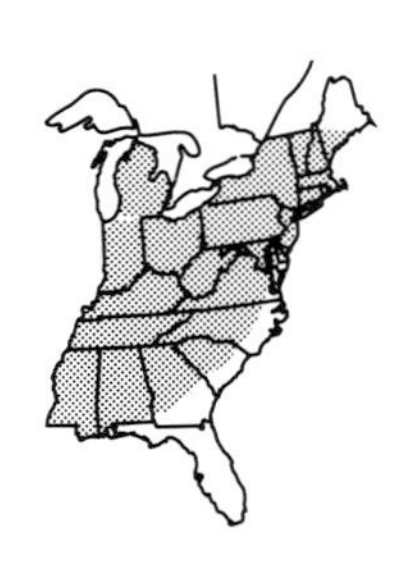

The pinnate leaves of Box Elder may be yellow, bronze, or brown, but are never red. Each leaf has 3–7 elliptic or ovate, serrate leaflets 2.5–5 cm (1–2") long. Pendant stalks bearing paired flat, light brown "keys" or fruits remain on the twigs through the fall and help in the identification of Box Elder. This fast-growing, often weedy tree is found in lowland forests and along forest margins, along roadsides, and in waste places.

Although the leaves of mature trees are usually a drab tan or greenish gray in autumn, the yellow leaves of stump sprouts can be quite attractive.

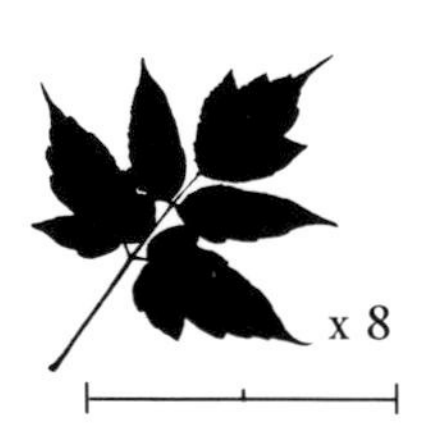

T-O/POS Maple Family (*Aceraceae*)

61 a

61 b

62

Please Be Careful

Do not spoil your enjoyment of the fall woodlands by acquiring a most uncomfortable case of dermatitis from one of our deceptively colorful, but to some people very poisonous, native plants! Learn to recognize the three leaflets and white berries of the weedy, common, and widespread vines of Poison Oak/Poison Ivy and the pinnate leaves and white berries of Poison Sumac (#66), a less common but colorful small tree found in swamps and bogs in eastern North America.

> UNLESS YOU KNOW YOU ARE IMMUNE, DO NOT TOUCH THESE PLANTS! IF YOU DO, WASH THOROUGHLY WITH SOAP AND WATER AT ONCE.

63. Poison Oak **AVOID!**

Rhus toxicodendron

64. Poison Ivy

Rhus radicans

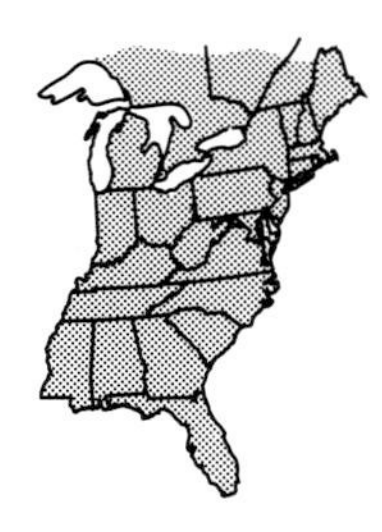

The three weakly lobed leaflets of Poison Oak and the three unlobed leaflets of Poison Ivy are 5–20 cm (2–8") long and are so variable in size, shape, and fall color that there is no good line of separation between these two plants. They do, however, have one very important characteristic in common: all parts of these vines (of both leaf forms or species), the roots, stems, leaves, and berries, contain an oil that even in very minute amounts can cause painful blisters on the skin of any susceptible person.

The white berries (#112) are eaten and widely dispersed by birds. These weedy, semi-shrubby, prostrate or high-climbing vines occur in low woodlands, along streams and trails, and anywhere there is a clearing: in old fields, along fencerows and roadsides, and in weedy areas.

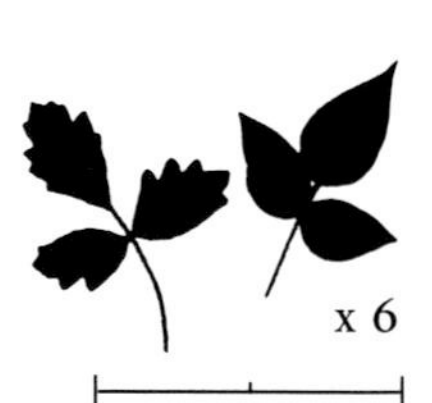

V-A/TOE Cashew Family (*Anacardiaceae*)

63 a

63 b

64

65. Fragrant Sumac

Rhus aromatica

The three weakly lobed leaflets of this fragrant shrub look very much like those of Poison Oak, but the leaflets are generally smaller, 2.5–8 cm (1–3") long, and thicker, and Fragrant Sumac, which has red fruits in midsummer, is not poisonous. Also note that next year's flower buds are present at the tip of the branches when the leaves turn color in the fall.

These shrubs, often found on rocky limestone soils (or granite outcrops!), are sporadic over much of the eastern United States. Because of their fragrance and fall color, they are used in landscaping.

S-A/TOL Cashew Family (*Anacardiaceae*)

66. Poison Sumac AVOID!

Rhus vernix

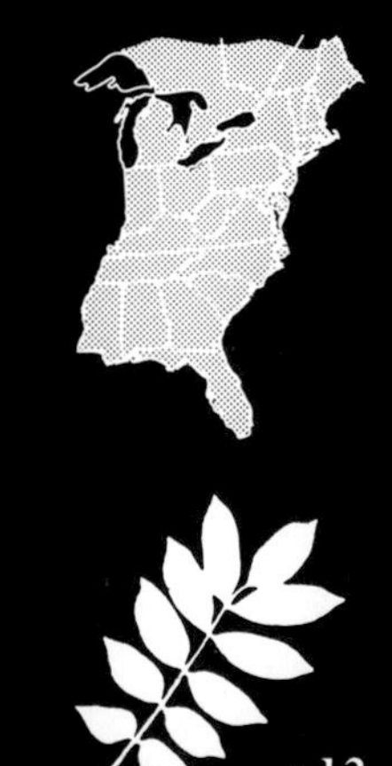

The brilliant yellow, orange, or red leaves of Poison Sumac are pinnate and have 7–13 elliptic to oblong leaflets 5–12 cm (2–5") long, with smooth margins. All parts of this colorful small tree are poisonous to those not immune to the irritating oil.

Poison Sumac usually grows in swamps, bogs, and low moist areas, often in the shade of taller trees. Fortunately, it is not as weedy and common as Poison Oak and Poison Ivy, but the oil is stronger and produces a more painful irritation.

T-A/PEE Cashew Family (*Anacardiaceae*)

65

66 a

66 b

67. Winged Sumac
Rhus copallina

The glossy maroon leaves of this common, widespread (and nonpoisonous) shrub or small tree are pinnate, usually with 9–11 elliptic, entire leaflets 3–8 cm (ca. 1–3") long that appear to be connected by a narrow wing along the leaf stalk. These plants are rhizomatous and often form large clumps on the dry open banks of our newer highways.

The conical cluster of small fruits, which are as colorful as the fall leaves, can be soaked in water to make a yellowish tea or "lemonade." The thick, soft brown stems have been used as a source of tannin.

x 7

S-A/PEE Cashew Family (*Anacardiaceae*)

68. Staghorn Sumac
Rhus typhina

The large dark crimson — or rarely orange to yellow — pinnate leaves of this sumac have 15–31 narrow, elliptic to lanceolate, serrate leaflets that can be up to 15 cm (6") long. The stems and fruits of Staghorn Sumac are densely hairy, as in the "velvet stage" of a stag's antlers.

These colorful, nonpoisonous shrubs or small trees are rhizomatous. The large clumps form bright splashes of color along forest margins, road banks, and fencerows and in old fields over much of the Appalachians. Staghorn Sumac is rare or absent at lower elevations in the south.

x 18

S-A/PNS Cashew Family (*Anacardiaceae*)

67 a [BGH]

67 b [FCS]

68

69 a

69. Smooth Sumac
Rhus glabra

The long, pinnate, scarlet leaves of Smooth Sumac are similar to those of Staghorn Sumac, but are glabrous rather than hairy or pubescent. The bright red pubescent fruits of this nonpoisonous shrub are effective in the landscape or in dried arrangements.

Smooth Sumac is a common, colorful rhizomatous shrub of eastern North America that grows in a variety of open habitats at both lower and higher elevations. Large, colorful colonies often occur along our newer highways.

S-A/PNS Cashew Family (*Anacardiaceae*)

69 b [FCS]

70. Hercules Club
Aralia spinosa

The bronze leaflets of these large, twice compound, often spiny leaves, the large compact clusters of small, black, fleshy fruits (#123), and the sharp spines on the stout trunks identify these shrubs or small trees. Hercules Club is found as a native in moist woods and roadsides of the southern Appalachian area, but is spreading and becoming naturalized northward.

Use of the aromatic root as a toothache remedy in Colonial times gave this spiny tree another name: Toothache Tree.

T-A/BNS Ginseng Family (*Araliaceae*)

71. Witch Alder
Fothergilla major

Witch Alder's maroon to bright red, orange, or yellow leaves are 5–10 cm (2–4") long, are ovate to elliptic or obovate, and are weakly toothed toward the apex.

Although rare in the dry woods and on the high balds of the extreme southern Appalachians, these showy native shrubs often form large colonies in more favorable habitats. They have attractive flowers in the spring and are widely used for landscaping in the northeastern states.

S-A/SEI Witch Hazel Family (*Hamamelidaceae*)

71 [PSW]

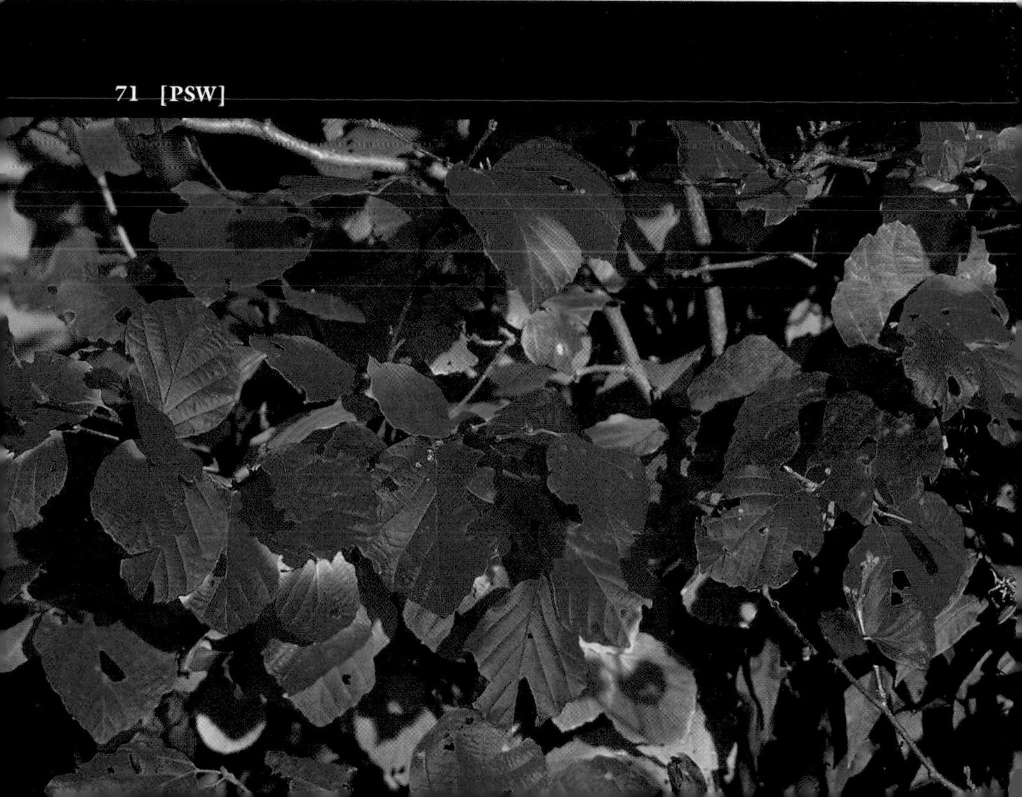

72 a

72. Sweetgum
Liquidambar styraciflua

The autumn colors of the star-shaped leaves of Sweetgum range from very dark maroon to brilliant red, soft pink, orange, or clear yellow. The leaves are 7–15 cm (ca. 3–6") long, usually have 5 pointed lobes, and appear to have more intense fall color if the trees are growing in poor soil. The pendant, round, spiny fruits, about 2.5 cm (1") in diameter (#131), are quite distinctive. Also the bark of young twigs may have corky wings or ridges.

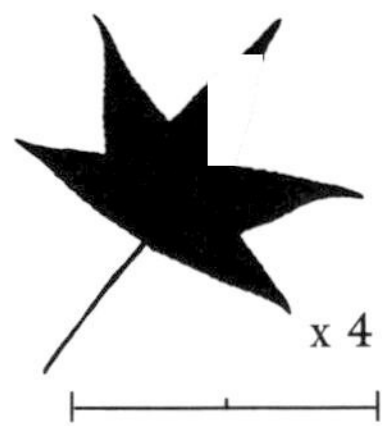

This colorful and important timber tree grows to 30 m (100') and is common in low woods and along stream banks at lower elevations, generally below 800 m (2,500'), from southern New England to Florida. The wood is used for furniture, veneer, plywood, and pulpwood; it burns well but is very difficult to split. The sweet gum that exudes from cuts in the bark was used medicinally in Colonial times.

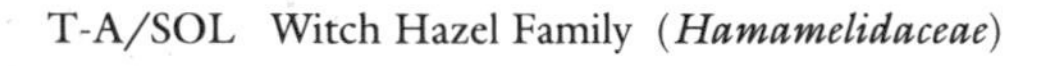

T-A/SOL Witch Hazel Family (*Hamamelidaceae*)

72 b

72 c

72 d

72 e

73. Virginia Creeper

Parthenocissus quinquefolia

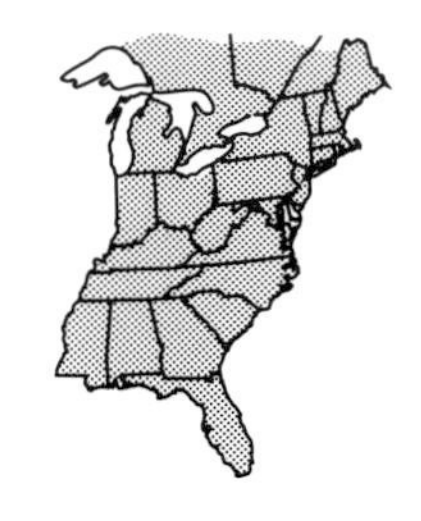

The 5 leaflets of the palmately compound leaves of the nonpoisonous Virginia Creeper and its dark blue fruits easily distinguish it from Poison Ivy, which has 3 leaflets and white fruit. Virginia Creeper's ovate to elliptic, serrate leaflets are up to 15 cm (6") long. Compare the pictures here with those of Poison Oak and Poison Ivy (#63, #64) to be sure you can tell these woody vines apart!

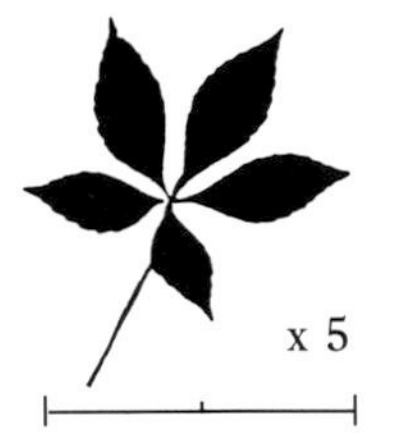

Virginia Creeper establishes quickly in clearings and along fencerows, stone walls, roadsides, and forest margins over much of eastern North America. It is colorful and prominent in the fall, either on the ground or up in the trees, especially along some of our interstate highways and in many coastal areas of New England.

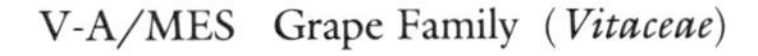

V-A/MES Grape Family (*Vitaceae*)

74. Greenbriar

Smilax glauca

The large, bright red leaves of Smilax or Greenbriar are spaced along the slender green stems of these tough, thorny perennial vines and often festoon the branches of small trees in open woodlands and roadside thickets from the Poconos to Florida. The alternate leaves vary in outline from almost round to triangular, are 5–10 cm (2–4") long, and have smooth margins.

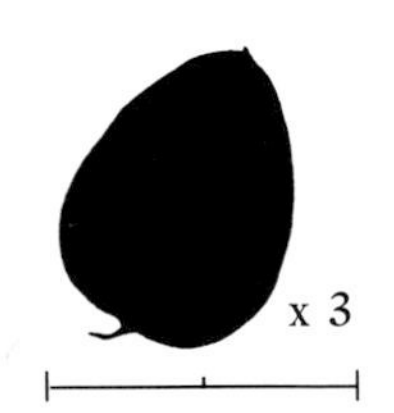

Wildlife eat the dark blue berries of *Smilax* species (#110 and #111) and spread the seeds. The enlarged rhizomes of some species were used by the Indians as a source of starch.

V-A/SRE Lily Family (*Liliaceae*)

73 a

73 b

74

75 a

75. Blackgum

Nyssa sylvatica

The bright red leaves of this tall tree, also called Tupelo, are among the most brilliant in the fall forest. The leaves are 5–13 cm (2–5") long and vary from elliptic to oblong or slightly obovate; the margin is entire, but a few leaves may have 1–2 small teeth. Blackgum grows in pine or deciduous forests on a variety of moist soils over the eastern United States. The related Water Tupelo (*Nyssa aquatica*) grows in swamps of the southern coastal plain.

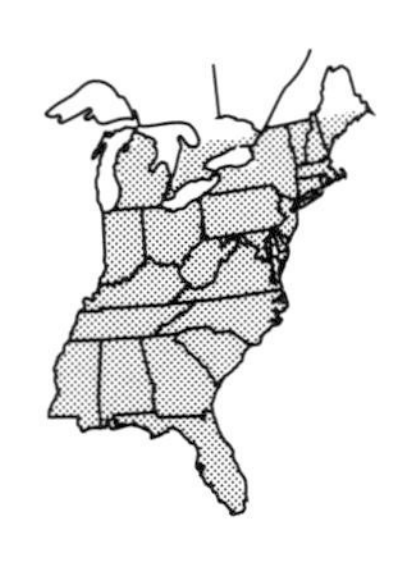

Gum trees are attractive ornamentals, and bees produce an excellent honey from their nectar. The wood is hard to split for fuel, and lumber from these trees is of poor quality and used only for crates, pallets, and similar rough items. The fleshy fruits (#124) are eaten by both birds and mammals.

T-A/SEE Gum Family (*Nyssaceae*)

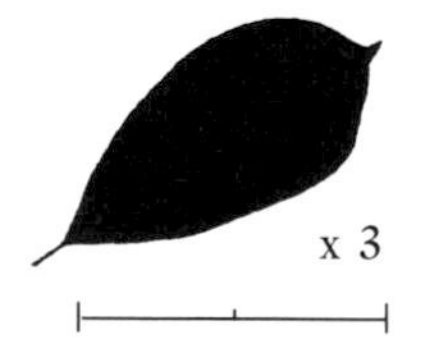

75 b

76. Blueberry

Vaccinium vacillans

The glossy maroon or red leaves of plants belonging to a dozen or so different coastal and mountain species of *Vaccinium* may provide no more than a small splash of color from a single bush. Alternatively, plants of rhizomatous species may form large colonies and produce extensive color, especially in burned or cleared areas — such as Graveyard Fields along the Blue Ridge Parkway — where Blueberry is the primary plant cover. The leaves of all Blueberries are relatively small, 2–8 cm (ca. 1–3") long, finely serrate, and generally deciduous (see figure 76 b, *Vaccinium constablaei*).

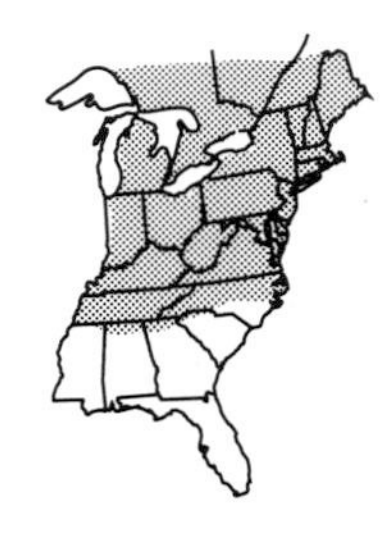

Our native Blueberries, which furnished the ancestral stock of our prolific commercial varieties, are still enjoyed by campers, hikers, and, of course, birds, bears, and other woodland animals.

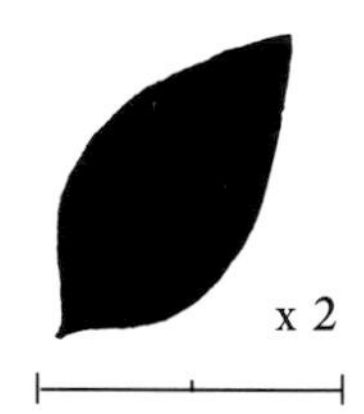

x 2

S-A/SES Heath Family (*Ericaceae*)

77. Buckberry

Gaylussacia ursina

The bright pink to red elliptic leaves of Buckberry — or Bearberry, as it is called by some — are a common fall sight in and around the Great Smoky Mountains National Park. These endemic, slender, rhizomatous understory shrubs literally cover the ground under the oaks and Mountain Laurel at middle and high elevations in their limited range in the southern Appalachians. The leaves are 5–10 cm (2–4") long and, unlike the leaves of Blueberries, have small resinous glands on the undersurface.

The black berries, which contain 10 small, hard seeds, are usually eaten by birds and other forest animals as soon as they are ripe. People gather what is left.

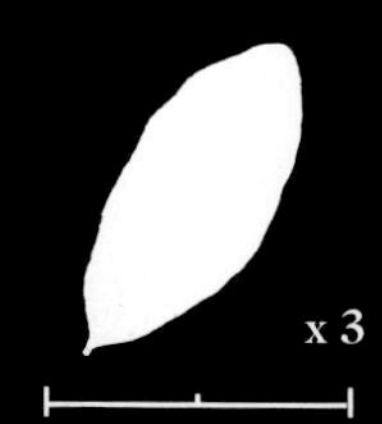

x 3

S-A/SEE Heath Family (*Ericaceae*)

76 a

76 b

77

78. Sourwood
Oxydendrum arboreum

By late August it is not uncommon to see the foliage of young Sourwood trees along the sides of the Blue Ridge Parkway beginning to turn red. The process will continue well into October, at which time these trees are dark to bright red with flat sprays of small green fruits providing a colorful contrast. The elliptic, finely serrate leaves, which may be yellow in heavy shade, are 8–20 cm (ca. 3–8") long and have a sour taste.

Sourwood honey is often available where the trees are plentiful in the southern Appalachians. The trunks of smaller trees (about 15 cm or 6" in diameter), which are often sharply angled, were used for the runners of the mule-drawn land sleds that moved material from the outside world into the mountains before the wagon roads were built.

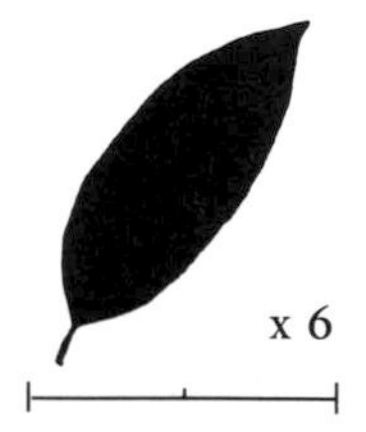

x 6

T-A/SES Heath Family (*Ericaceae*)

79. Fetterbush
Leucothoe recurva

The bright crimson fall foliage of the deciduous Fetterbush stands in brilliant contrast to the dark evergreen leaves of nearby Rhododendron, Mountain Laurel, and Dog-hobble (*Leucothoe axillaris*). The elliptic to lanceolate, serrate leaves are 5–13 cm (2–5") long. These twiggy shrubs may be 1–4 m (3–12') tall. They grow in bogs or in rocky woods at higher elevations in the southern Appalachians, where this colorful shrub is endemic.

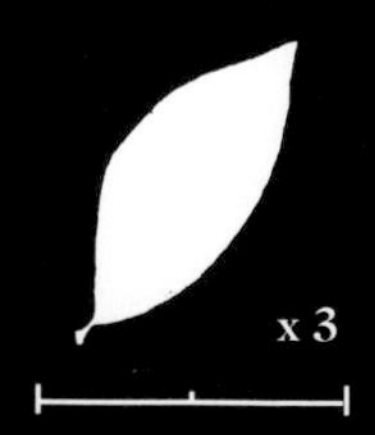

x 3

S-A/SES Heath Family (*Ericaceae*)

78 a

78 b [BGH]

79

80. Flowering Dogwood
Cornus florida

The fleshy berries of this small tree begin to turn red weeks before the opposite, elliptic leaves, 6–10 cm (2.5–4") long, take on their fall shades of red to maroon, or rarely yellow. Flowering Dogwood is a familiar understory tree of deciduous woodlands throughout most of the eastern United States.

The small size, attractive white bracts or "flowers," and colorful berries (#118) and leaves of Flowering Dogwood make it an important landscape plant. The hard wood from the small trunks was used in the past to make farm implements, wedges to split rails, and shuttles for spinning mills.

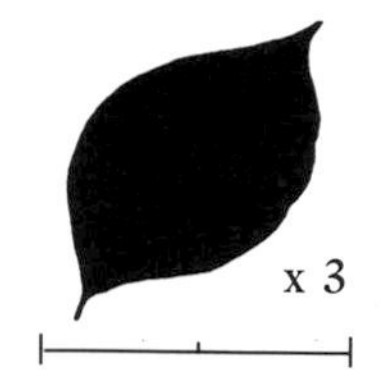

T-O/SEE Dogwood Family (*Cornaceae*)

81. Gray Dogwood
Cornus racemosa

The opposite elliptic leaves of these shrubs are red to maroon on the upper surface and greenish gray beneath; they are 4–8 cm (1.5–3") long. The gray to white fruits and the gray twigs help account for the common name. Gray Dogwood often forms large clumps in moist soils of meadows and roadsides in the northern Appalachian area, where it is far more abundant than it is in the mountains to the south.

The Roughleaf Dogwood (*Cornus drummondii*), which also has white fruits, is quite similar. As its name implies, however, its leaves are rough. Also it has a very different range, being most common in the midwest.

80 a

80 b

81

82. Downy Arrowwood
Viburnum rafinesquianum

The rose to maroon, ovate, opposite leaves of this attractive understory shrub are 4–8 cm (1.5–3") long, shallowly serrate, and downy pubescent beneath. The common name comes from this last characteristic plus the reported Indian use of the straight stems for arrow shafts. These shrubs, which grow to 2 m (6') tall, are a prominent component of drier hardwood forests throughout the Appalachians and adjacent foothills.

The black fruits (#116), which taste a bit like raisins when dried, are quickly eaten by wildlife.

S-O/SOS Honeysuckle Family (*Caprifoliaceae*)

83. Blackhaw
Viburnum prunifolium

The autumn leaves of Blackhaw range from orange-red for woodland plants to deepest maroon for plants growing in full sun. The glossy, elliptical, opposite leaves are 2.5–7.5 cm (1–3") long and are finely serrate. The primary range of these shrubs or small trees, which grow in bogs and low, moist woodlands at intermediate elevations, is from the mid-Atlantic states westward to Missouri. The fleshy black fruits have a hard, flattened pit or stone and are readily eaten by both birds and mammals.

S-O/SES Honeysuckle Family (*Caprifoliaceae*)

82

83 a

83 b [FCS]

84. Maple-leaf Viburnum

Viburnum acerifolium

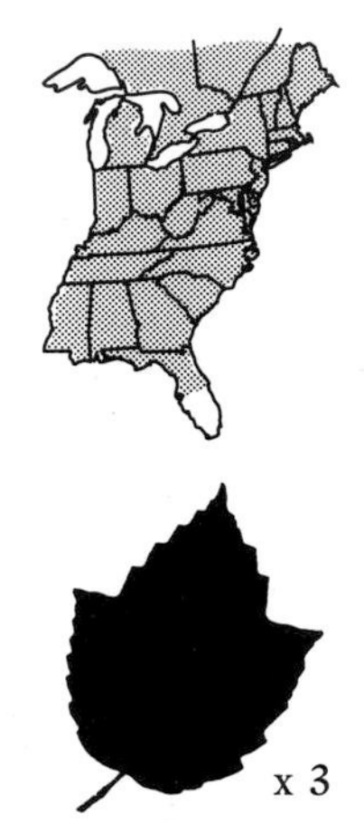

The pinkish to orange-red or maroon, serrate, sharply 3-lobed, opposite leaves of Maple-leaf Viburnum are 5–13 cm (2–5") long and do indeed look much like maple leaves. These low shrubs are rhizomatous and form large clumps or colonies in deciduous woodlands over much of eastern North America.

As with other shrubs of this genus, the fleshy, purple-black fruits are an important food for wildlife. The decorative plants are also used in landscaping.

S-O/SOL Honeysuckle Family (*Caprifoliaceae*)

85. Squashberry

Viburnum edule

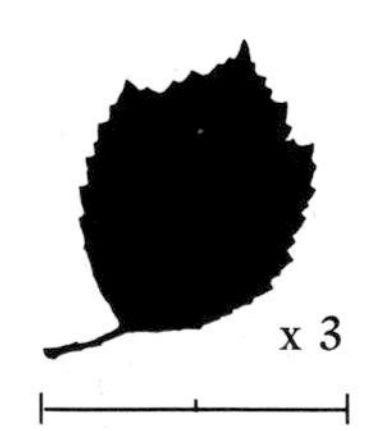

The maroon, 3-lobed leaves of Squashberry, 5–10 cm (2–4") long and sharply serrate, resemble those of Maple-leaf Viburnum, but the fleshy fruits of this shrub are bright red or orange and squash easily. Squashberry, another plant of moist northern woodlands and clearings, grows from Labrador to Alaska. It is rare or absent south of the Adirondacks.

The juicy fruit, eaten by birds and other wildlife, is also collected for making jelly.

S-O/SOL Honeysuckle Family (*Caprifoliaceae*)

84

85

86

86. Hobblebush
Viburnum alnifolium

The foliage of Hobblebush begins to turn color by midsummer. The bright hues of yellow, orange, red, or maroon of the large, rounded, paired leaves makes a spectacular splash of color, especially when the plants occur with evergreens, in the northern forests of the Appalachians and on southward to the Smokies. The distinctive leaves are 10–20 cm (4–8") long and nearly as wide, with finely serrate margins and rusty pubescence beneath.

The fleshy fruits, bright red before they ripen to black, are eaten by wildlife. Indeed, as the common name Moosewood indicates, the plants may be browsed by moose and also by deer. Uncontrolled browsing by deer can lead to the plant's extinction in a given area.

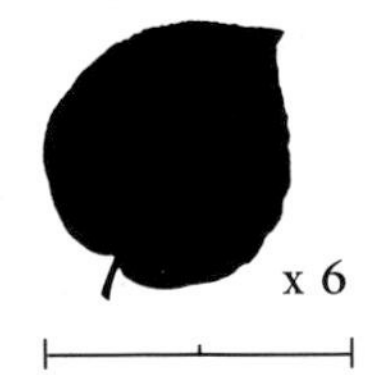

S-O/SRS Honeysuckle Family (*Caprifoliaceae*)

With a dozen species represented, the oaks are the largest and most varied group of trees in our list. Their dark reds and browns, and a few yellows, are often major components of the fall color over much of the southern Appalachians.

Since oak leaves are high in tannins, the leaves of a given tree may turn directly from green to brown, or a single leaf may have some green, some red, and some brown areas, depending upon its internal chemistry. The reds of oak leaves range from dull to rich to brilliant, depending on the species and the environment. Yellows are infrequent, and an entire oak with good yellow foliage is an exception.

Botanically, the oaks are divided into two major groups: the white oaks and the black (or red) oaks. In the white oaks the acorns mature in one season and the lobes of the leaves are rounded and smooth (no bristle tips). In the black oaks, the acorns mature in the fall of their second year and the leaves have more angular and bristle-tipped lobes. The oaks are wind-pollinated in early spring before the leaves are mature, and the different species within each group hybridize freely, often making identification difficult.

The oaks, with perhaps 80 native species, are by far the most important group of hardwood timber trees in North America. The strong, durable, white wood of the white oaks is in great demand for furniture, flooring, paneling, casks and barrels, construction timbers, fuel, and many other uses. A cord (128 cubic feet) of dried white oak is equivalent to a ton of coal in heat value. Thin, strong, flexible white oak "splits" are still used to weave baskets, chair seats, and fish traps. Black oak wood is of slightly lower quality; it is not used to make casks for aging bourbon whiskey, but otherwise used much like the wood of white oaks.

The oaks as a group are important landscape plants. The vast shower of acorns each fall is a major part of the mast crop on which the Indians and early settlers depended, indirectly, for meat and which sustains much of the forest wildlife today.

87 a

87. White Oak
Quercus alba

White Oak leaves are a rather dull red, brown, bronze, or yellow, the color slow to form and often not uniform on a single leaf, branch, or tree. The leaves are 10–25 cm (4–10") long, with 5–9 rounded lobes and a smooth margin. Some trees may hold their brown, dead leaves until late fall or even early winter. These large, important timber trees grow to 33 m (100') or more tall and are frequent in upland woods throughout their extensive range.

The acorns of White Oak (#146) provide an important food for squirrels, wild turkeys, and other animals, and at one time were also used for food by the Indians.

White Oak is the state tree of Connecticut, Maryland, and Illinois.

T-A/SEL Beech Family (*Fagaceae*)

87 b

87 c

87 d

88 a

88 b

88. Chestnut Oak
Quercus prinus

The leaves of Chestnut Oak are usually brown to bronze-yellow, but may also be dull red. They are 10–20 cm (4–8") long and easily identified by the numerous (15–25) shallow, rounded lobes along the smooth margin. The leaves of the closely related Swamp Chestnut Oak (*Q. michauxii*, figure 88 c), which has a more coastal distribution, are definitely obovate and pubescent beneath. These large trees, which grow to about 25 m (75'), may be widely scattered or in pure stands on dry or rocky upland soils of the Appalachian region.

The large acorns of Chestnut Oak, 2–3 cm long (figure 88 b, top; and #144), are an important game food (note size of *Q. stellata*, center, and *Q. velutina*, bottom). The strong, white wood, very similar to that of White Oak, generally has the same uses. The bark was an important source of tannin in earlier times.

T-A/SEL Beech Family (*Fagaceae*)

88 c

89. Post Oak
Quercus stellata

The leaves, 10–15 cm (4–6") long, may be dull or lustrous red, bronze, or yellow, but are more often mottled brown and green. This medium-sized tree grows to 12–15 m (40–50') on rocky, dry, or sandy slopes and ridges. On poor soils, it may be more shrub-like and only half normal height.

The small acorns (figure 88 b, center; and #145) add to the mast crop. The hard wood, as with other members of the white oak group, is valuable for mine timbers, railroad ties, fence posts, barrels, flooring, and firewood.

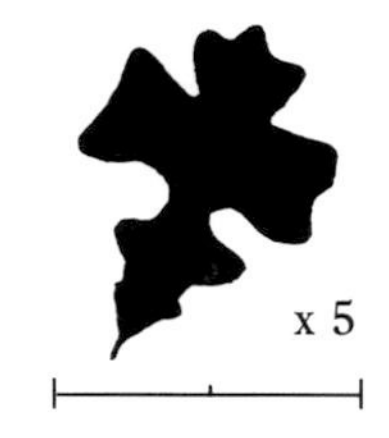

T-A/SBL Beech Family (*Fagaceae*)

90. Red Oak
Quercus rubra

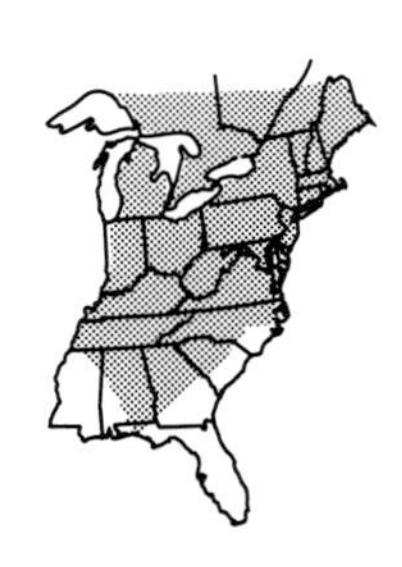

Red Oak leaves are round to obovate in general outline, may vary from red to greenish yellow or brown, and are 12–20 cm (5–8") long. They have 7–11 rather triangular, bristle-tipped lobes with more open sinuses, or spaces between the lobes, than Scarlet Oak. Red Oak is a fairly common but variable species that grows to 25 m (75') on rich soils in the mountains. It is less frequent at lower elevations. The large acorns, about 2.5 cm (1") long, have a very shallow "cup" around the base (#147).

The darker wood of Red Oak has many of the same uses as White Oak, such as flooring, shelving, furniture, and fuel.

Red Oak is the state tree of New Jersey.

T-A/SBL Beech Family (*Fagaceae*)

89

90

91 a

91 b

91. Black Oak

Quercus velutina

The large red, reddish brown, or yellow-brown leaves of Black Oak are 15–20 cm (6–8") long and roughly ovate to obovate in general outline, with 5–9 bristle-tipped lobes; they are smooth on both surfaces by fall. These rather common trees grow to 20 m (60') or so in well-drained soils of slopes and ridges at lower elevations over much of eastern North America.

The acorns are about 2 cm (.75") long, with the bottom one-third covered by the scaly cup (see figure 88 b, bottom, for size comparison). However, fruit set is variable and sporadic, and some years no acorns form. The bark was once a source of tannin and also, in Colonial times, provided a yellow dye.

T-A/SOL Beech Family (*Fagaceae*))

92 [BGH]

92. Scarlet Oak

Quercus coccinea

Few trees in the fall landscape can match the brilliance of Scarlet Oak. The leaves all turn scarlet about the same time and often stay on the tree until they dry and turn brown late in the fall. The widely elliptic leaves are 10–18 cm (4–7") long and have 5–9 narrow, angular, bristle-tipped lobes. This medium-sized tree reaches a height of about 25 m (75') on relatively poor, dry soils throughout its range.

The lumber from this oak is of poorer quality than that from other members of the black oak group, but the wood is hard and makes good fuel. Also, because of its striking fall color, Scarlet Oak is used in landscaping.

T-A/SEL Beech Family (*Fagaceae*)

93. Pin Oak
Quercus palustris

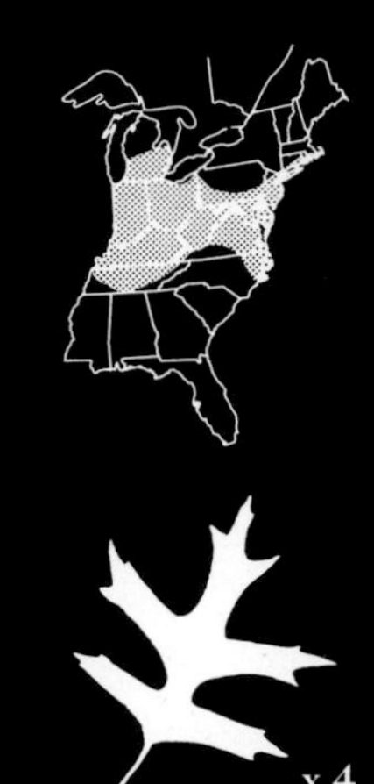

The red or brownish red, widely ovate leaves of Pin Oak are 5–15 cm (2–6") long, with 5–7 angular, usually 3-pointed, lobes with bristle tips; they are smooth except for tufts of hairs in the axils of the veins on the lower surface. Pin Oak trees grow to 25 m (75') or taller in bottomlands and moist low grounds. The small acorns are 1–2 cm (ca. .5") long.

As with the other black oaks, the wood is strong, is straight-grained, and splits easily into shingles or into the slender "pins" or "pegs" formerly used instead of nails to fasten the timbers of buildings. Pin Oak is frequently used in landscaping.

x 4

T-A/SOL Beech Family (*Fagaceae*)

94. Willow Oak
Quercus phellos

The brown to golden brown, linear, unlobed leaves of Willow Oak are 5–10 cm (2–4") long and have a single bristle at the tip, which identifies the tree as a member of the black oak group. This large tree, 25–30 m (75–90') or taller and with a crown almost as wide, grows in moist lowland soils of the southern coastal plain.

The small acorns, which are often borne in abundance, are about 1 cm (ca. .5") long and have a shallow cup. The wood of Willow Oak is of poorer quality than that of many other oaks, but the tree grows rapidly and is widely used as an ornamental or street tree within its range.

x 3

T-A/SLE Beech Family (*Fagaceae*)

93 a

93 b

94 a

94 b

95. Blackjack Oak
Quercus marilandica

The broad, glossy, bristle-tipped, bluntly 3-lobed leaves of Blackjack Oak may turn bronze but often go directly from green to brown. These small crooked or irregular trees, 7–15 m (20–50') tall, grow in rather poor dry upland soils in much of the southeast.

The elongate, elliptical acorns are 1.5–2 cm (ca. .75") long and are half-enclosed by the deep cup. The wood is used for fuel, charcoal, rough timbers, and railroad ties.

T-A/SBL Beech Family (*Fagaceae*)

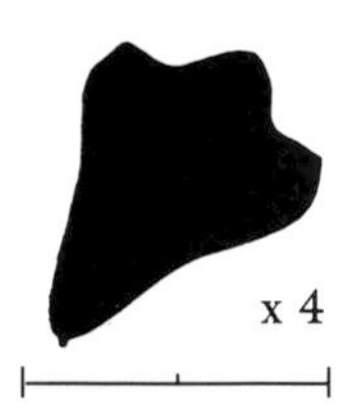

x 4

96. Water Oak
Quercus nigra

In the fall the foliage of Water Oak may remain green until late in the season and then turn brown and drop from the tree, or some of the leaves may turn golden yellow. The typically obovate or spatulate leaves are 5–10 cm (2–4") long. This coastal plain tree grows rapidly to 20 m (60') or more on moist but well-drained soils.

The small black or dark brown acorns are almost round and are only 1–1.5 cm (ca. .5") long. Some are striped with orange and are quite colorful. The wood can be used for fuel or for second-quality lumber. The tree is often used in landscaping in its range.

x 3

95

96

97. Turkey Oak
Quercus laevis

Turkey Oak has brilliant scarlet leaves, 10–20 cm (4–8") long, with 3–5 angular, bristle-tipped lobes and tufts of brownish hairs in the angles of the veins on the lower surface. The leaves are slow to drop and retain their striking color for several weeks. These small, gnarled, often shrub-like trees are frequent in the southern sandhills and coastal plain. They may be 6–10 m (18–30') or taller. Since they regenerate after fire, a small tree may have a very old and very large root system.

The ovoid or egg-shaped acorns are about 2.5 cm (1") long. Fruit production in Turkey Oak is often sporadic, but large stands of these trees may be produced asexually from underground tissue.

T-A/SOL Beech Family (*Fagaceae*))

98. Spanish Oak
Quercus falcata

The yellow-green to golden brown leaves of Southern Red Oak, as this species is also called, are 10–20 cm (4–8") long with 3–5 narrow, bristle-tipped lobes; the long, narrow end lobe is very distinctive. These trees may be 15–25 m (45–75') tall and are frequent on dry, rather poor soils of the southern piedmont and parts of the coastal plain.

The acorns are small, about 1–1.5 cm (ca. .5") long, with the bottom one-third or less covered by a shallow cup. The wood is sold as red oak, and the tree is often used in landscaping.

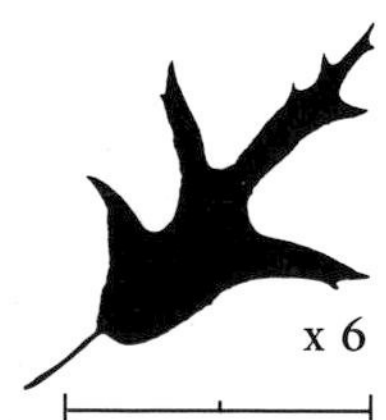

T-A/SOL Beech Family (*Fagaceae*)

97 a

97 b

98

99 a

99 b

99. Sycamore
Platanus occidentalis

The pale gray to bright white bark of the upper trunk and spreading limbs of Sycamore stands out along the courses of streams and the edges of moist floodplains on the brown winter landscape, more than making up for the usual sedate brown tones of the tree's autumn foliage. The broad, 3-lobed, sharply serrate leaves are usually 10–20 cm (4–8") long and wide. The base of the trunk of these trees may be up to 1 m (3') or more in diameter; a historic Sycamore along the Ohio River had a reported circumference of 14 m (47'). The bark on the lower trunk is dark gray and rough.

These large, widespread trees are an important source of timber. In Colonial times, the cores of the round fruits (#129), along with the tough fruit stalk and a short piece of the twig, were used for buttons.

T-A/SCL Sycamore Family (*Platanaceae*)

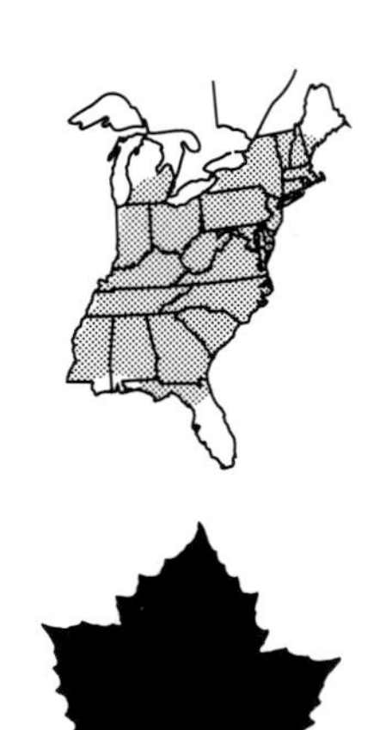

99 c
[GCP]

100 a

100. Fraser Magnolia

Magnolia fraseri

The large brown, or rarely yellow, leaves of Fraser Magnolia are alternate, but they are so crowded at the ends of the stout twigs that they appear to be in a whorl. This is true also of Umbrella Magnolia (*Magnolia tripetala*), which is also found in the southern Appalachians but which has larger leaves without the "ears" or basal lobes.

An interesting pattern of concentric circles can sometimes be found (by a close observer) on leaves of Fraser Magnolia that have been infected by some forest-dwelling microorganism.

T-A/SBL Magnolia Family (*Magnoliaceae*)

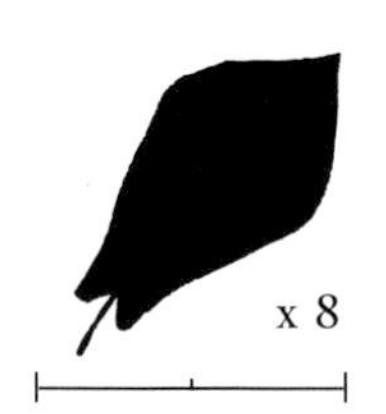

100 b
[BGH]

Woodland Harvests: Fruits and Seeds

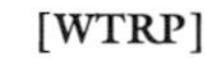

[WTRP]

Woodland Harvests

When people think of woodland harvests today, they think of such things as nuts, berries, roots, and perhaps a few mushrooms that can be gathered for fall or winter use; or perhaps of the roots, leaves, and fall flowers that are collected and dried to make natural dyes for yarn. Seldom does anyone think of the important annual harvests of wood that is cut for fuel, or the meadows that are mowed for hay for livestock. Nor do most people remember that corn, pumpkins, and cranberries are all native American plants and that apples were introduced very early into the New World from Europe and soon became naturalized. Thus even though these crops have been vastly improved by the plant breeders and even though they are harvested and transported by modern machines and equip-

ment, our traditional autumn harvest of corn, pumpkins, and apples is, historically, a harvest from the American woodlands.

Although the woodland harvest can provide many fruits, seeds, and roots for human use or consumption, the word "harvest" is best understood as the *total* production of stored energy of all kinds, both food and fuel, by the plants of our fields and forests. This includes the production of many fruits, seeds, roots, and leaves that are not used by man but are an essential part of nature's vast and beautifully integrated food chain. Thus the great majority of nature's bounty provides the food (called "mast") for wildlife, a category that includes birds, mammals, opossums, reptiles, and even insects and worms.

A Note on the Kinds of Fruits

In a truly mature hardwood or pine forest, very little light penetrates the dense canopy to reach the forest floor. Since the light is so restricted, it is not surprising that there are very few understory plants of any sort and there are no seedlings. Even if light were not a limiting factor, any seedling that sprouted under its mature and well-established parent tree would be doomed. Throughout the biological world the fierce competition for survival dictates that offspring cannot compete directly with their parents. In addition, continued establishment of offspring near a parent ultimately promotes inbreeding, which generally lowers the viability of an organism in the struggle for existence. Accordingly, throughout the world of plants, a very diverse and interesting array of mechanisms have developed that effect the dispersal of fruits and seeds.

Generally, fruits are divided into two major groups: fleshy and dry. Beyond this first division, because of the rich diversity of fruit structure, form, and function, a standardized classification of fruit types becomes too difficult and technical for a book like this. Thus in the following pages the fleshy fruits, which are primarily eaten by birds and are often brightly colored, are divided only by the growth form of the parent plant: herb, vine, shrub, or tree. The dry fruits, which are usually brown or gray, are divided by size into two broad categories: those that are very small and eaten whole by seed-eating birds and small mammals; and those that are fairly large and contain dry seeds eaten by larger birds and mammals. Some dry fruits (or seeds in the case of Milkweed and Indian Cigar) are winged and are wind-dispersed; these are kept together. Another large class of dry fruits, nuts, are also kept together.

Although some of the species included in the Woodland Harvests section were treated earlier (and are cross-referenced by entry number), many plants that do not have bright fall foliage but that have colorful fruits have been added. Because the emphasis of this book is on fall col[illegible]umber of interesting but relatively drab burrs, seeds, and small fruits are not illustrated.

101. Jack-in-the-Pulpit
Arisaema triphyllum
(Araceae)
Bogs and rich woods; throughout.

102. Doll's Eyes
Actaea pachypoda
(Ranunculaceae)
Rich northern woods; south in the mountains.

101

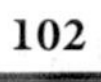

102

103

104

105 [WSJ]

103. Painted Trillium

Trillium undulatum
(*Liliaceae*)
Moist northern woods; southward in the mountains.

104. Ginseng

Panax quinquefolium
(*Araliaceae*)
Rich woods; south in the mountains.

105. False Solomon's Seal

Smilacina racemosa
(*Liliaceae*)
Rich woods; throughout.

106. Wintergreen

Gaultheria procumbens
(*Ericaceae*)
Moderately dry hardwood forests; south in the mountains.

107

108 [WSJ]

107. Horse Gentian
Triosteum angustifolium
(Caprifoliaceae)
Deciduous woods on basic soil; southern Appalachians.

108. Cranberry
Vaccinium macrocarpon
(Ericaceae)
Northern bogs; rare in high southern mountains.

109. Bittersweet (#29)
Celastrus scandens
(Celastraceae)
Rich soil of roadsides and thickets; more or less throughout.

110. Carrion Flower
Smilax herbacea
(Liliaceae)
Moist, open woodlands and thickets; throughout.

109 [FCS]

111

112

113

111. Greenbriar (#74)

Smilax glauca
(Liliaceae)
Alluvial woods, old fields; throughout southern Appalachians.

112. Poison Ivy (#64)

Rhus radicans
(Anacardiaceae)
Essentially ubiquitous!

113. Strawberry Bush

Euonymus americanus
(Celastraceae)
Moist woods; throughout southern Appalachians.

114. Silverberry

Elaeagnus angustifolia
(Elaeagnaceae)
Naturalized along roadsides at scattered localities; throughout.

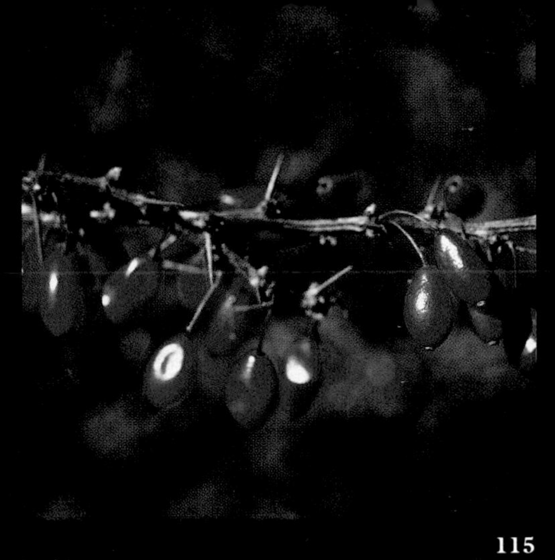

115

116

115. Barberry
Berberis thungbergii
(Berberiaceae)
Naturalized in open areas on limestone soils; infrequent.

116. Downy Arrowwood (#82)
Viburnum rafinesquianum
(Caprifoliaceae)
Upland hardwood forests, throughout the Appalachian area.

117. Possum Haw
Ilex decidua
(Aquifoliaceae)
Upland woods and thickets; southern Appalachian area.

118. Flowering Dogwood (#80)
Cornus florida
(Cornaceae)
Deciduous and evergreen woodlands; throughout.

117

118 [BGH]

119

120

119. Hawthorn

Crataegus sp.
(Rosaceae)

One or more species are found in thickets and woodlands; throughout.

120. Red Cedar (#6)

Juniperus virginiana
(Cupressaceae)

Woods, old fields, and fencerows on limestone soils; throughout.

121. Persimmon

Diospyros virginiana
(Ebenaceae)

Woodlands and old fields; southern New England to Florida.

The fleshy fruits of Persimmon are excellent to eat when they are fully ripe, but they are hard and astringent when green. A fermented mash of persimmon pulp and Honey Locust pods makes "locust beer," and the sweet pulp makes an excellent persimmon pudding. These fruits are also a favorite with wildlife from birds to bears.

121 [BGH]

122

123

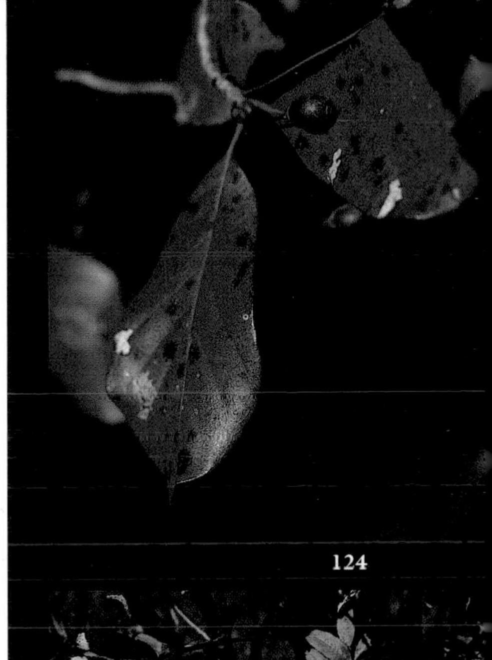

124

122. Osage Orange

Maclura pomifera
(Moraceae)

Sparingly naturalized (from the midwest) in low pastures and fencerows.

123. Hercules Club (#70)

Aralia spinosa
(Araliaceae)

Moist roadsides and wet woods; southern Appalachians.

124. Blackgum (#75)

Nyssa sylvatica
(Nyssaceae)

Moist deciduous woods; essentially throughout.

125. Mountain Ash (#55)

Sorbus americana
(Rosaceae)

Rocky outcrops; south along the mountains.

126. Crabapple

Malus coronaria

(Rosaceae)

Woodland borders in the southern Appalachians.

Tree Fruits Seen Against the Sky

Five of our native or naturalized trees can often be easily identified, even from a distance, long after all of their leaves have fallen because they have distinctive fruits that may hang on the twigs and branches well into the winter. This special series of fruit photographs was kindly made by David H. Ramsey.

127. Indian Cigar

Catalpa speciosa

(Bignoniaceae)

Naturalized (from further west) in waste areas, pastures, and woodland margins at scattered localities in the southern Appalachian area. The slender cylindrical fruits are 30 cm (1') or more long. The numerous flat, winged seeds in each capsule are 3–4 cm (ca. 1.5") long and are wind-dispersed.

127 [DHR]

128. Princess Tree

Paulownia tomentosa
(Scrophulariaceae)

Naturalized (from China) in waste places, in open areas, and along roadsides of the southern Appalachian area. The "candelabra" of round, tomentose brown buds will produce large purple flowers next spring. The round, open capsules of this year's blooms are empty but still evident; each one produced hundreds of minute, round, flattened seeds with a marginal wing that were dispersed by the wind.

129. Sycamore (#99)

Platanus occidentalis
(Platanaceae)

Alluvial woods and streambanks more or less throughout. The round fruits, about 2.5 cm (1") in diameter, are made up of many small, hard seeds, each with a plume of brown hairs that aid in wind dispersal.

130. Honey Locust

Gleditsia triacanthos
(Fabaceae)

Floodplains and waste places, often on limestone at lower elevations generally west of the crest of the Appalachians. The broad pods, 30 cm (1') or more long, contain a dozen or more hard, flat beans embedded in a sweet pulp. The pods are used, with persimmons, in locust beer and are also eaten by animals.

128 [DHR]

129 [DHR]

130 [DHR]

131 [DHR]

131. Sweetgum (#72)
Liquidambar styraciflua
(Hamamelidaceae)

Moist soils with other hardwoods below 1,000 m (3,000') elevation in the southern Appalachian area. The spiny fruits are about 2.5 cm (1") in diameter and have two winged seeds in each of the many capsules.

132. Ironwood (#40)

Carpinus caroliniana
(Betulaceae)

Rich alluvial soils at elevations below 1,000 m (3,000') throughout eastern North America. The 3-lobed wing on each fruit aids in wind dispersal.

133. Tulip Poplar (#43)

Liriodendron tulipifera
(Magnoliaceae)

Moist deciduous or pine woods, essentially throughout the eastern United States. The flat, winged fruits are spread by the wind.

134. Green Ash (#57)

Fraxinus pennsylvanica
(Oleaceae)

Moist woodlands and stream banks at lower elevations throughout the eastern United States. The slender, flat, winged fruits break from the cluster individually and are spread by the wind.

132 [BGH]

133

135

136

135. Silverbell

Halesia carolina
(Styracaceae)

Understory tree in moist hardwood forests of the southern Appalachians. The narrow wings on the fruit aid somewhat in dispersal by the wind.

136. Milkweed

Asclepias sp.
(Asclepiadaceae)

One or more species of these herbaceous perennials occur in fields, in clearings, and along roadsides throughout eastern North America. The plumose seeds can be very widely dispersed by the wind.

137. Lily

Lilium superbum
(Liliaceae)

Infrequent in wet meadows and open, moist woodland borders over eastern North America. The light, flat seeds may be blown short distances by the wind.

137

138. Cattail

Typha latifolia
(Typhaceae)

Common in wet ditches, pond margins, and marshes throughout eastern North America. The thousands of minute, plumose brown fruits that make up the "cattail" are widely dispersed by the wind.

139. Virgin's Bower

Clematis virginiana
(Ranunculaceae)

Moist soils of road banks, thickets, and forest margins over eastern North America. The plumose fruits are wind-dispersed.

140. Black Walnut (#48)

Juglans nigra
(Juglandaceae)

Now rare in low woods over much of the eastern and central United States. The round fruits, or nuts, may be spread by squirrels or by gravity or water.

138

139

141

142

141. Mockernut Hickory (#51)

Carya tomentosa
(Juglandaceae)

Upland pine-oak-hickory forests over much of the eastern United States. The hard nuts are spread by squirrels.

142. Buckeye (POISON !)

Aesculus sp.
(Hippocastanaceae)

One or more of our 4 species occur in rich mountains or low-elevation woods from Pennsylvania south. The thin-shelled nuts, about 2.5 cm (1") in diameter, may roll downhill or may be dispersed by squirrels.

143. Chinquapin

Castanea pumila
(Fagaceae)

On road banks and in forest clearings at scattered localities in the southern Appalachians. Once the small, sweet nuts, about 1 cm (3/8") in diameter, drop out of the "burrs," they may be spread by various small mammals and large birds.

144. Chestnut Oak (#88)

Quercus prinus
(Fagaceae)

Rocky slopes and ridges of the Appalachian area.

144

145. Post Oak (#89)

Quercus stellata
(Fagaceae)

In a variety of dry to moist habitats throughout the southeast.

146. White Oak (#87)

Quercus alba
(Fagaceae)

Well-drained soils more or less throughout the eastern United States.

147. Red Oak (#90)

Quercus rubra
(Fagaceae)

Rich soils, uplands of the southern Appalachian area.

Acorns may be moved downslope by gravity or water, but most distribution is probably by various birds and mammals (and especially squirrels) who store or eat the nutritious nuts.

145

146

Glossary, Keys, Appendixes, Index

Glossary

Apex. Tip; the end of the leaf away from the stem.

Appressed. Flattened, pressed against, not upright.

Catkin. A pendant, usually flexuous, elongate cluster of small, often unisexual, flowers.

Colonial. A plant that can spread by runners, rhizomes, root shoots, or other asexual means.

Compound leaf. A leaf made up of separate leaflets.

Deciduous. Trees or shrubs that shed all their leaves each year; not evergreen.

Dioecious. Having the male and female reproductive organs on separate plants.

Doubly serrate. With small teeth along the margins of the larger teeth.

Ecotone. Where two different habitats meet, as a field and a forest.

Endemic. Restricted to a relatively small area or region.

Entire. A margin without teeth, lobes, or divisions.

Glabrous. The surface smooth and without trichomes or hairs.

Lateral veins. The secondary side veins of a leaf.

Leaflet. A single unit or division of a compound leaf which will ultimately separate from the leaf axis by an abscission layer.

Nutlets. Very small, hard, one-seeded fruits.

Oblique. Slanting; unequal-sided.

Perennial. Plant of three or more years' duration.

Pith. Soft central portion of a stem.

Pubescent. Covered with short, soft trichomes or plant hairs.

Rhizome. An underground stem often giving rise to new shoots.

Serrate. With sharp teeth pointing forward.

Spatulate. Oblong, but tapered to the base, spoon-like.

Stellate. Star-shaped.

Tannin. A brown chemical compound in leaves and bark used to tan leather.

Tomentose. Densely woolly or pubescent; with matted, soft, wool-like trichomes or plant hairs.

Leaf Identification Keys

(For the Plant Groups Illustrated)

Note: The entry number (not the page number) is given for each species, genus, or group of species. Where 2 or 3 different species key out together at the end of a line, they can be easily separated and identified by their respective illustrations.

If the leaves are:

Green	Yellow	Red	Brown
go to	go to	go to	go to
G-1	Y-1	R-1	B-1

Leaves Green

G-1a Low, herbaceous plants; leaves elongate, pinnately lobed .... Ferns (#14–15)

G-1b Trees or shrubs, leaves slender, needle-like, single or in bundles of 2–5, or small and scale-like; trees .... go to G-2

G-1c Leaves wide, flat, leathery; not needle- or scale-like; trees or shrubs .... Broad-leaf Evergreens

Leaves over 6 cm (2.5") long
- Tree, coastal .... Magnolia (#13)
- Shrub, mountain .... Rhododendron (#10)

Leaves less than 6 cm (2.5") long
- Leaves spiny, tree .... Holly (#12)
- Leaves not spiny, shrub .... Mountain Laurel (#11)

G-2a Trees with needles

Needles single
- Needles flat .... Fir (#1), Hemlock (#2)
- Needles square, stiff .... Spruce (#3)

Needles in bundles
- Five needles per bundle .... White Pine (#7)
- Two needles per bundle .... Jack Pine (#8), Virginia Pine (#9)

G-2b Trees or shrubs with leaves scale-like
- Leaves in flattened sprays; cones woody .... White Cedar (#4)
- Leaves not in flattened sprays; cones fleshy ... Red Cedar (#6), Creeping Cedar (#5)

Leaves Red

R-1a Leaves simple .... go to R-2

R-1b Leaves compound, made up of 3 or more leaflets .... go to R-5

R-2a Margins smooth, leaf not lobed .... go to R-3

R-2b Margins toothed or leaf lobed .... go to R-4

R-3a Plant a low shrub, or a vine

Shrub .... Blueberry (#76)
Buckberry (#77)

Vine .... Greenbriar (#74)

R-3b Plant a tree

Leaves opposite .... Dogwoods (#80,81)

Leaves alternate

Some leaves 2–3 lobed .... Sassafras (#58)

Leaves all unlobed .... Blackgum (#75)

R-4a Margins toothed

Leaves alternate; trees or shrubs

Trees .... Sourwood (#78)
Cherry (#26)

Shrubs .... Blueberry (#76)
Buckberry (#77)
Fetterbush (#79)

Leaves opposite; shrubs .... Haws (#82–86)

R-4b Margins lobed, leaves opposite

Trees .... Maples (#59–60)

Shrubs .... Haws (#82–86)

R-4c Margins lobed, leaves alternate

Shrub .... Witch Alder (#71)

Tree

Leaves star-shaped .... Sweetgum (#72)

Leaves not star-shaped .... Oaks

Leaves widest beyond middle .... Blackjack Oak (#95)
Water Oak (#96)

Leaves usually widest at middle

Leaves with shallow or rounded lobes, not bristle-tipped

Lobes shallow .... Chestnut Oak (#88)

Lobes deeply cut .... White Oak (#87)

Leaves with deeply cut, pointed lobes, bristle-tipped

Plant of coastal sandhills .... Turkey Oak (#97)

Plant of inland soils .... Red Oak (#90)
Scarlet Oak (#92)
Pin Oak (#93)

R-5a Leaves palmate; leaflets 3 or 5 .... go to R-6

R-5b Leaves pinnate; leaflets more than 5

Leaves opposite .... Ash (#56)
Leaves alternate
  Stem spiny; leaves twice comound .... Hercules Club (#70)
  Stem smooth; leaves once compound ..... Sumacs (#66–69)

R-6a Leaflets 5 .... Va. Creeper (#73)

R-6b Leaflets 3

Shrub; fruits red .... Fragrant Sumac (#65)
Vine; fruits white .... Poison Oak/Ivy (#63–64)

Leaves Yellow

Y-1a Leaves simple .... go to Y-2

Y-1b Leaves compound .... go to Y-8

Y-2a Leaves unlobed, margins smooth .... go to Y-3

Y-2b Leaves lobed or margins toothed .... go to Y-5

Y-3a Plant a vine

Leaves round; fruit fleshy, red or yellow .... Bittersweet (#29)
Leaves cordate; fruit dry, brown .... Pipe Vine (#22)

Y-3b Plant a tree or shrub .... go to Y-4

Y-4a Plant a shrub

Leaves alternate .... Spicebush (#20)
Leaves opposite
  Leaves and stems fragrant; fruit dry, brown .... Sweet Shrub (#24)
  Leaves and stems not fragrant; fruit fleshy, dark blue .... Fringe Tree (#23)

Y-4b Plant a tree

Leaves in tufts, linear .... Larch (#25)
Leaves opposite, elliptic .... Fringe Tree (#23)
Leaves alternate
  Leaves linear .... Willow Oak (#94)
  Leaves cordate .... Redbud (#21)
  Leaves elliptic .... Pawpaw (#19)
  Cucumber Magnolia (#18)

Y-5a Leaves strongly or deeply lobed

Vines .... Grapes (#30–31)
Trees
  Leaves opposite .... Maples (#59–62)

Leaves alternate
Leaf broadly notched at tip .... Tulip Poplar (#43)
Leaf star-shaped .... Sweetgum (#72)
Leaf margin serrate .... Mulberries (#32–33)
Leaf margin entire
Some leaves unlobed.............. Sassafras (#58)
All leaves lobed
Lobes bristle-tipped .... Spanish Oak (#98)
Black Oak (#91)
Lobes not bristle-tipped .... White Oak (#87)
Chestnut Oak (#88)

Y-5b Leaves with shallow lobes or margins toothed ...... go to Y-6

Y-6a Shrub or vine; leaves about as wide as long

Shrub; leaf base oblique .... Witch Hazel (#28)
Vine; leaf base cordate .... Grapes (#30–31)

Y-6b Trees; leaves twice or more as long as wide .... go to Y-7

Y-7a Bark of tree rough, brown or dark

Lateral veins not conspicuous, not strongly parallel, not ending in a marginal tooth ...... Mulberries (#32–33)
Lateral veins conspicuous, parallel, ending in a marginal tooth
Leaf base oblique, upper or lower surface rough or some twigs with corky ridges .. Elms (#44–46)
Leaf base not oblique, leaf surfaces smooth
Leaves more than 12.5 cm (5")......... Chestnut (#47)
Leaves less than 12.5 cm (5")
Twigs pungent or aromatic .... Cherries (#26–27)
Sweet Birch (#37)
Twigs not scented .... Hop Hornbeam (#39)

Y-7b Bark thin, smooth (often peeling), white, gray, tan, or yellow
Bark peeling or twigs aromatic .... Birches (#35–38)
Bark not peeling, twigs not aromatic
Leaves triangular or ovate, lateral veins curved, not parallel .... Aspens (#41–42)
Leaves elliptic, lateral veins straight, parallel .... Beech (#34)
Ironwood (#40)

Y-8a Leaflets 3, palmate .... Poison Oak/Ivy (#63–64)

Y-8b Leaflets more than 3, pinnate .... go to Y-9

Y-9a Plants herbaceous; generally 1–1.5 m (3–4.5') or less tall .... Ferns (#16–17)

Leaves opposite .... Ash (#56)
Leaves alternate
Stem spiny; leaves twice compound .... Hercules Club (#70)
Stem smooth; leaves once compound ..... Sumacs (#66–69)

Y-9b Plant woody, trees or shrubs more than 1.5 m (4.5') tall .... go to Y-10

Y-10a Shrubs; fruit red or maroon

Stem with thorns .... Rose (#54)
Stem without thorns
Stem thick, pith large .... Sumacs (#67–69)
Stem not thick, pith small .... Mountain Ash (#55)

Y-10b Trees; fruit brown or black

Leaves opposite .... Ashes (#56–57)
Leaves alternate
Stems with thorns .... Locust (#53)
Stems without thorns .... Hickories (#49–52)
Walnut (#48)

Leaves Brown

B-1a Leaves without lobes .... Cucumber Magnolia (#18)
Beech (#34)

B-1b Leaves lobed

Lobes only 2, basal .... Fraser Magnolia (#100)
Lobes 3-5 or more
Lobes palmate, pointed .... Sycamore (#99)
Lobes pinnate, rounded .... White Oak (#87)
Post Oak (#89)

Appendix A
Sources of Fall Color Information

For information on the best fall color routes, fall color schedules, and travel information for specific areas, a brief letter or telephone call to one or more of the following three offices will produce an array of helpful free brochures and maps. These will, in turn, provide further names and addresses for more local information.

1. The state Office of Travel/Tourism of the state(s) you plan to visit. This is the easiest place to start, since many states now have a toll-free (800) visitor information telephone number. To get this number for a particular state (if they have one), call the 800 number directory service at 1-800-555-1212. Some states with 800 visitor information numbers have made them easy to remember, for example: 1-800-847-4862 (=1-800-VisitNC); 1-800-225-5982 (=1-800-CallWVa); 1-800-225-5697 (=1-800-CallNYS); and, for Ohio, 1-800-282-5393 (=1-800-Buckeye).

2. The local Chamber of Commerce of a town in the area you plan to visit. In many of the most enjoyable scenic areas population density is low, and so several communities may work together to provide visitor information on fall color and other seasonal events. Direct information from such local visitor bureaus is often quite specific and helpful.

3. The U.S. Forest Service Headquarters for any specific National Forest along your route.

In addition, the various state offices of a number of private travel and automobile associations, such as the American Automobile Association, often have special fall color maps, up-to-date foliage reports, and other helpful seasonal information for their members.

As with any popular seasonal event, your fall color trip will be more enjoyable if your plans and reservations are made as far in advance as practical.

Appendix B
Daylight/Temperature Chart

Average number of hours of sunlight* by month for three northern latitudes

	30°N Jacksonville	40°N Philadelphia	50°N N. Maine
Jan	10:25	9:40	8:20
Feb	11:11	10:42	10:07
Mar	11:48	11:53	11:51
Apr	12:53	13:13	13:47
May	13:39	14:23	15:23
Jun	14:00	14:59	16:21
Jul	13:54	14:44	15:57
Aug	13:14	13:46	14:29
Sep	12:11	12:28	12:38
Oct	11:27	11:11	10:47
Nov	10:39	9:59	9:03
Dec	10:14	9:21	8:06
6-month difference in sunlight hours	**3:46**	**5:38**	**8:15**
ave. minutes/day speed of change	**1:25**	**1:78**	**2:43**

***Daylight is longer**

Seasonal changes in day length (photoperiod) at three northern latitudes

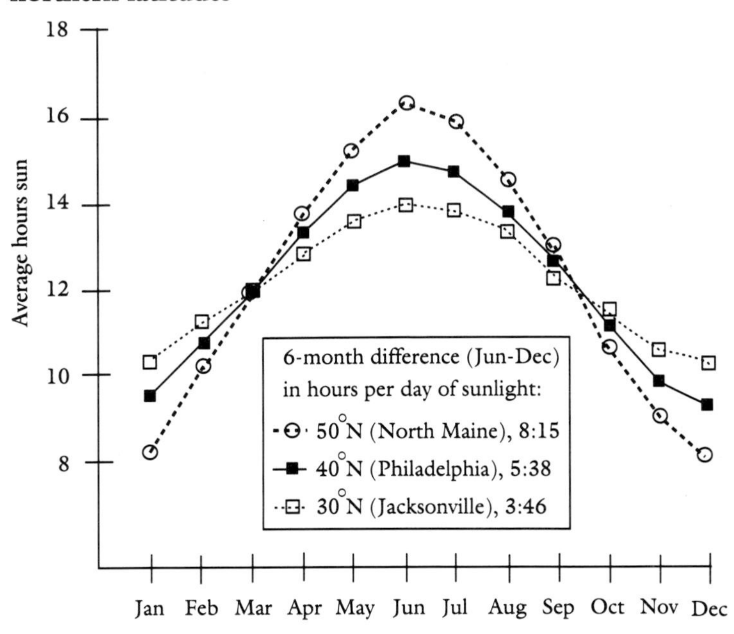

Average monthly temperatures (°F) for seven northern latitudes

Latitude City	34°N Atlan	36°N Knoxv	38°N Chltsv	40°N Phila	42°N Bngmtn	44°N Rtlnd	46°N Hoult
Jan	42	38	35	31	21	21	14
Feb	45	42	37	33	22	23	16
Mar	53	50	46	42	31	33	27
Apr	62	60	57	53	44	45	39
May	69	67	65	63	55	57	52
Jun	76	74	73	72	64	65	62
Jul	79	78	77	77	69	70	67
Aug	78	77	76	75	67	68	64
Sep	73	72	70	**68**	**60**	**60**	**56**
Oct	**62**	**60**	**59**	**57**	**49**	**50**	**45**
Nov	**52**	**49**	**49**	**46**	38	39	33
Dec	45	41	39	36	26	26	19

Fall color time! (shaded cells)

Index
(Numbers are Entry Numbers)